SigmaPlot® 2000
Programming Guide

Exact Graphs for Exact Science

For more information about SPSS® Science software products, please visit our WWW site at *http://www.spss.com* or contact

SPSS Science Marketing Department
SPSS Inc.
233 South Wacker Drive, 11th Floor
Chicago, IL 60606-6307
Tel: (312) 651-3000
Fax: (312) 651-3668

Contents

Contents

Introduction

The *Programming Guide* provides you with complete descriptions of SigmaPlot's powerful math, data manipulation, regression, and curve fitting features. It also describes how to use SigmaPlot's Interactive Development Environment (IDE) and Macro Recorder to automate and customize SigmaPlot tasks.

Transforms

Transforms are sets of equations that manipulate and calculate data. Math transforms apply math functions to existing data and also generate serial and random data. To perform a transform, you enter variables and standard arithmetic and logic operators into a *transform dialog*. Your equations can specify that a transform access data from a worksheet as well as save equation results to a worksheet.

Transforms can be saved as independent .XFM files for later opening or modification. Because transforms are saved as plain text (ASCII) files, they can be created and edited using any word processor that can edit and save text files.

The transform chapters describe the use and structure of transforms, followed by a brief tutorial, reference sections on transform operators and functions, and finally a list and description of the sample transform files and graphs included with SigmaPlot.

Regressions

The SigmaPlot *Regression Wizard* replaces the older curve fitter with a new interface and over one hundred new equations. The major new features of this interface include:

➤ a graphical interface rather than text code
➤ a library of over 100 built-in equations in twelve different categories
➤ graphical examples of the curves and equations for built-in equations

➤ automatic initial parameter determination—no coding is required in most cases

➤ selection of variables directly from either worksheet columns or graph curves

➤ full statistical report generation

➤ automatic curve plotting to existing or new graphs

➤ new regression equation documents for the notebook

➤ new text report documents for the notebook

The Regression Wizard chapters describe how to use these features.

The Curve Fitter

The Regression Wizard uses the *curve fitter* to fit user-defined linear equations to data. The curve fitter modifies the parameters (coefficients) of your equation, and finds the parameters which cause the equation to most closely fit your data.

You can specify up to 25 equation parameters and ten independent equation variables. When you enter your equation, you can specify up to 25 parameter constraints, which limit the search area when the curve fitter checks for parameter values.

The curve fitter can also use weighted least squares for greater accuracy.

User-defined equations can be saved to notebooks or regression libraries and selected for later use or modification.

Automation

SigmaPlot *OLE Automation* technology provides you with a wide range of possibilities for automating frequently-performed tasks, using macros and user-defined features.

SigmaPlot's *Macro Recorder* lets you record is a set of procedures and then run them automatically with a single command. Most of the operations that you perform in SigmaPlot can be recorded.

The Macro Window provides a fully-featured programming environment that uses *SigmaPlot Basic* as the core programming language. If you are familiar with Microsoft Visual Basic, most of what you know will apply as you use SigmaPlot's macro language.

2

Using Transforms

Transforms are math functions and equations that generate and are applied to worksheet data. Transforms provide extremely flexible data manipulation, allowing powerful mathematical calculations to be performed on specific sets of your data.

Using the Transform Dialog Box

To begin a transform, choose the Data menu User Defined Transform command or press F10. The User-Defined Transform dialog box appears.

Figure 2–1
The User-Defined
Transform Dialog Box

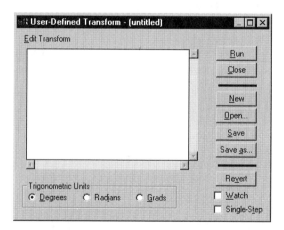

Creating a Transform

The first step to transform worksheet data is to enter the desired equations in the edit box. If no previously entered transform equations exist, the edit box is empty: otherwise, the last transform entered appears.

Select the edit box to begin entering transform instructions. As you enter text into the transform edit box, the box scrolls down to accommodate additional lines.

Up to 100 lines of equations can be entered. Equations can be entered on separate lines or on the same line.

Once you have completed the transform, you can run it by clicking Run.

Transform Files
Transforms can be saved as independent transform files. The default extension is .XFM. Transform files are plain text files that can also be edited with any word processing program.

Use the New, Open, Save, and Save As options in the User-Defined Transform dialog box to begin new transforms, open existing transforms, save the contents of the current edit box to a transform file, and save an existing transform file to a different file name.

The last transform you entered, opened, or imported always appears in the edit window when you open the User-Defined Transform dialog box. To permanently save a transform, you must use the Save, or Save As options.

Transform Syntax and Structure

Use standard syntax and equations when defining user-defined transforms in SigmaPlot or SigmaStat. This section discusses the basics and the details for entering transform equations.

Transform Syntax
Transforms are entered as equations with the results placed to the left of the equal sign (=) and the calculation placed to the right of the equal sign. Results can be defined as either variables (which can be used in other equations), or as the worksheet column or cells where results are to be placed.

Entering Transforms
To type an equation in the transform edit box, click in the edit box and begin typing. When you complete a line, press Enter to move the cursor to the first position on the next line.

Figure 2–2
Typing Equations
into the Edit Window

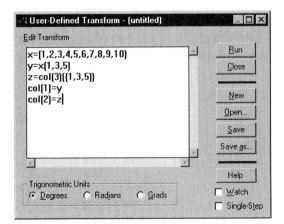

You can leave spaces between equation elements: $x = a+b$ is the same as $x = a + b$. However, you may find it necessary to conserve space by omitting spaces. Blank lines are ignored so that you can use them to separate or group equations for easier reading.

If the equation requires more than one line, you may want to begin the second and any subsequent lines indented a couple of spaces (press the space bar before typing the line). Although this is not necessary, indenting helps distinguish a continuing equation from a new one.

Σ You can resize the transform dialog box to enlarge the edit box. You can press Ctrl+X, Ctrl+C, and Ctrl+V to cut, copy, and paste text in the edit window.

Transforms are limited to a maximum of 100 lines. Note that you can enter more than one transform statement on a line; however, this is only recommended if space is a premium.

Σ Use only parentheses to enclose expressions. Curly brackets and square brackets are reserved for other uses.

Commenting on Equations

To enter a comment, type an apostrophe (') or a semicolon (;), then type the comment to the right of the apostrophe or semicolon. If the comment requires more than one line, repeat the apostrophe or semicolon on each line before continuing the comment.

Sequence of Expression

SigmaPlot and SigmaStat generally solve equations regardless of their sequence in the transform edit box. However, the col function (which returns the values in a worksheet column) depends on the sequence of the equations, as shown in the following example.

Example: The sequence of the equations:

col(1)=col(4)^alpha
col(2)=col(1)*theta

must occur as shown. The second equation depends on the data produced by the first. Reversing the order produces different results. To avoid this sequence problem, assign variables to the results of the computation, then equate the variables to columns:

x=col(4)
y=x^alpha
z=y*theta
col(1)=y
col(2)=z

The sequence of the equations is now unimportant.

Transform Components

Transform equations consist of *variables* and *functions*. *Operators* are used to define variables or apply functions to *scalars* and *ranges*. A scalar is a single worksheet cell, number, missing value, or text string. A range is a worksheet column or group of scalars.

Variables

You can define variables for use in other equations within a transform. Variable definition uses the following form

variable = expression

Variable names must begin with a letter: after that, they can include any letter or number, or the underscore character (_). Variable names are case sensitive—an "A" is not the equivalent of an "a." Once a variable has been defined by means of an expression, that variable cannot be redefined within the same transform.

Functions

A function is similar to a variable, except that it refers to a general expression, not a specific one, and thus requires arguments. The syntax for a function declaration is

function(argument 1,argument 2,...) = expression

where *function* is the name of the function, and one or more argument names are enclosed in parentheses. Function and argument names must follow the same rules as variable names.

User-Defined Functions   Frequently used functions can be copied to the Clipboard and pasted into the transform window.

Constructs

Transform constructs are special structures that allow more complex procedures than functions. Constructs begin with an opening condition statement, followed by one or more transform equations, and end with a closing statement. The available constructs are for loops and if...then...else statements.

Operators

A complete set of arithmetic, relational, and logic operators are provided. Arithmetic operators perform simple math between numbers. Relational operators define limits and conditions between numbers, variables, and equations. Logic operators set simple conditions for if statements. For a list of the operators and their functions, see Chapter "Transform Operators", on page 17.

Numbers

Numbers can be entered as integers, in floating point style, or in scientific notation. All numbers are stored with 15 figures of significance. Use a minus sign in front of the number to signify a negative value.

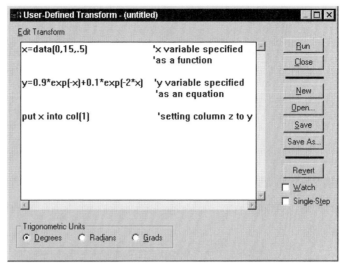

Missing values, represented in the worksheet as a pair of dashes, are considered non-numeric. All arithmetic operations which include a missing value result in another missing value.

To generate a missing value, divide zero by zero

Example: If you define:

missing = 0/0

the operation:

size({1,2,3,missing})

returns a value of 4.0. (The *size* function returns the number of elements in a range, including labels and missing values.)

The transform language does not recognize two successive dashes; for example, the string {1,2,3,--} is not recognized as a valid range. Dashes are used to represent missing values in the worksheet only.

Strings, such as text labels placed in worksheet cells, are also non-numeric information. To define a text string in a transform, enclose it with double quotation marks.

As with missing values, strings may not be operated upon, but are propagated through an operation. The exception is for relational operators, which make a lexical comparison of the strings, and return true or false results accordingly.

Scalars and Ranges

The transform language recognizes two kinds of elements: *scalars* and *ranges*. A scalar is any single number, string, or missing value. Anything that can be placed in a single worksheet cell is a scalar.

A range (sometimes called a vector or list) is a one-dimensional array of one or more scalars. Columns in the worksheet are considered ranges.

Ranges can also be defined using curly bracket ({ }) notation. The range elements are listed in sequence inside the brackets, separated by commas. Most functions which accept scalars also accept ranges, unless specifically restricted. Typically, whatever a function does with a scalar, it does repeatedly for each entry in a range. A single function can operate on either a cell or an entire column.

Example 1: The entry:

{1,2,3,4,5}

produces a range of five values, from 1 through 5.

Example 2: The operation:

{col(1), col(2)}

concatenates columns 1 and 2 into a single range. Note that elements constituting a range need not be of the same type, i.e., numbers, labels and missing values.

Example 3: The entry:

{x,col(4)*3,1,sin(col(3))}

also produces a range.

Array References

Individual scalars can be accessed within a range by means of the square bracket ([]) constructor notation. If the bracket notation encloses a range, each entry in the enclosed range is used to access a scalar, resulting in a new range with the elements rearranged.

Example: For the range:

x = {1.4,3.7,3.3,4.8}

the notation:

x[3]

returns 3.3, the third element in the range. The notation:

x[{4,1,2}]

produces the range {4.8,1.4,3.7}. The constructor notation is not restricted to variables: any expression that produces a range can use this notation.

Example: The operation:

col(3)[2]

produces the same result as col(3,2,2), or cell(3,2). The notation:

{2,4,6,8}[3]

produces 6. If the value enclosed in the square brackets is also a range, a range consisting of the specified values is produced.

Example: The operation:

col(1)[{1,3,5}]

produces the first, third, and fifth elements of column 1.

Figure 2–4
Range and Array Reference
Operations Typed into
the User Defined
Transform Window

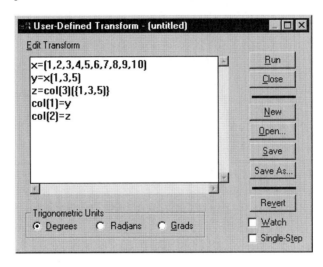

Notes

3

Transform Tutorial

The following tutorial is designed to familiarize you with some basic transform equation principles. You will enter transform data into a worksheet and generate a 2D graph.

Starting a Transform

To begin a transform:

1. Click the New Notebook ☐ button, or choose the File menu New command and select Notebook. An empty worksheet appears.

2. Choose the Data menu User-Defined Transform command. The User-Defined Transform dialog box appears. If necessary, click New to clear the edit window and begin a new session.

3. **Defining a Variable**   Click the upper left corner of the edit window and type:

   t=data(−10,11,1.5)

4. Add a few spaces, then type the comment:

   'generates serial data

   The data function is used to generate serial data from a specified start and stop, using an optional increment.

5. Press Enter to move to the next line, then type:

   col(1)=t 'put t into column 1

   This places the variable *t* into column 1 of the data worksheet.

6. Press Enter, then type:

cell(2,1) = "Results:" 'enclose strings in quotes

Figure 3–1
The Edit Window with
All the Transform
Equations Entered

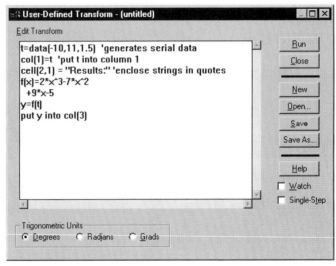

This places the label "Results:" in row one of column 2. Text strings must be enclosed in quotation marks.

7. **Defining a Function**   Press Enter, then type:

f(x)=2*x^3–7*x^2

Press Enter, add a couple of spaces, then type:

+9*x–5

If you want an equation to use more than one line, start each additional line with a blank space or two to distinguish it from a new equation.

8. Press Enter, then type:

   y=f(t)

   This variable declaration uses the function *f* and variable *t* declared in the previous equations.

9. Add a few spaces, then type:

   put y into col(3)

   This places the results of the preceding equation (which defines y) in column 3 of the worksheet.

   Note that you can also collapse the last two lines into one equation:

   col(3)=f(t)

   Click Run. If you have entered all the transform equations correctly, the data will appear as shown in Figure 3–2.

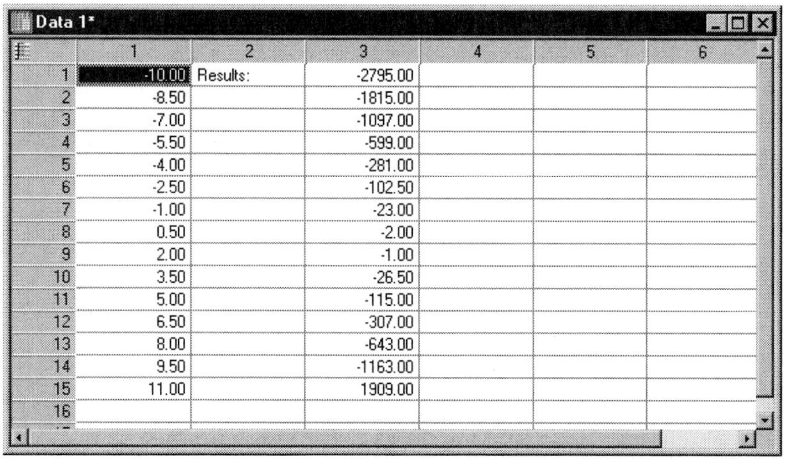

Figure 3–2
The Data Generated by
the Transform Tutorial

	1	2	3	4	5	6
1	-10.00	Results:	-2795.00			
2	-8.50		-1815.00			
3	-7.00		-1097.00			
4	-5.50		-599.00			
5	-4.00		-281.00			
6	-2.50		-102.50			
7	-1.00		-23.00			
8	0.50		-2.00			
9	2.00		-1.00			
10	3.50		-26.50			
11	5.00		-115.00			
12	6.50		-307.00			
13	8.00		-643.00			
14	9.50		-1163.00			
15	11.00		1909.00			
16						

Data 1*

Saving and Executing Transforms

After entering the transform equations, save the transform to a file, then run the transform.

1. Click Save, and specify a file name and destination for the file. The default extension for transform files is .XFM.

   Saved transforms can be opened with the Transform dialog box Open button.

2. Click Run. If you have entered all the transform equations correctly, you should generate the data shown in Figure 3–2.

Graphing the Transform Results

Once the transform is executed and the results are placed in the worksheet, you then treat the results like any other worksheet data.

1. Select a scatter graph from the graph toolbar and select a simple scatter graph.

   You can also choose the Graph menu Create Graph command, select Scatter Plot then click Next and select Simple Scatter.

2. Select XY Pair as the Data Format, then click Next. Select column 1 as your X column and column 3 as your Y column, then click Finish.

   A Scatter Plot graph appears. The data in column 1 is plotted along the X axis and the data in column 3 is plotted along the Y axis.

Figure 3–3
A Graph
of Plotting the
Transform Tutorial
Data as a Scatter Plot

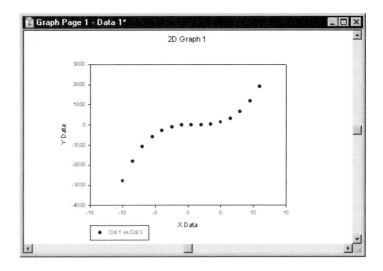

Recoding Example

This example illustrates a simple recoding transform.

1. Choose the Data menu User-Defined Transforms command to open the User-Defined Transform dialog box. If desired, click Save to save the existing transform to a file. Click New to begin a new transform.

Figure 3–4
Entering the Recoding
Transform Example
into the User-Defined
Transform Edit Window

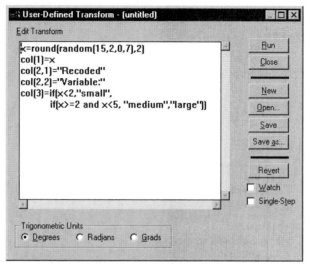

2. Click the upper left corner of the edit window and type:

    x = random(15,2,0,7)

    This creates uniformly random numbers distributed between 0 and 7, using 2 as a seed. However, the numbers generated have fifteen significant digits. To round off the numbers to two decimal places, modify this function to read:

    x = round(random,15,2,0,7),2)

3. Press Enter, then type:

    col(1) = x

    to place the random numbers in column 1.

4. Press Enter and type:

    col(2,1) = "Recoded "

Note the space between the d and the quotation mark ("). All characters, including space characters, within quotes are entered into cells as part of the label.

Press Enter, then type:

col(2,2) = "Variable:"

5. To create the code data, press Enter, then type:

col(3) = if(x<2,"small",

Press Enter, add a couple of spaces, then type:

if(x >=2 and x<5, "medium","large"))

If you want an equation to use more than one line, start each additional line with a blank space or two to distinguish it from a new equation.

6. Click Run. If you have entered all the transform equations correctly, the data will appear as shown in Figure 3–5.

Figure 3–5
Results of the Recoding
Example Transform

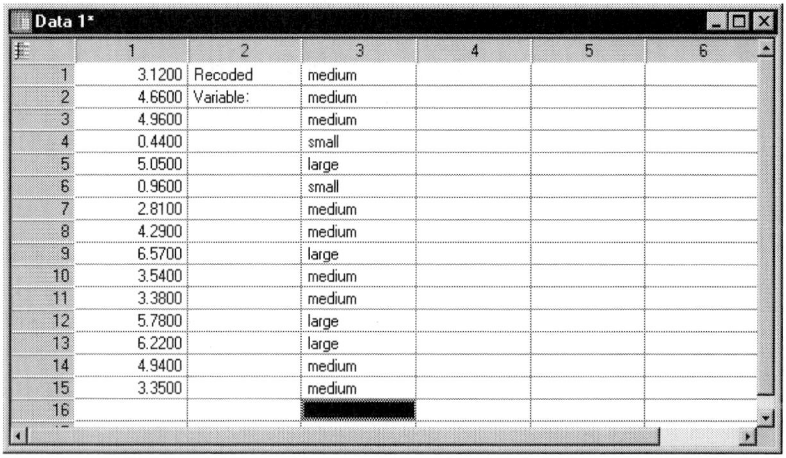

Data 1*						
	1	2	3	4	5	6
1	3.1200	Recoded	medium			
2	4.6600	Variable:	medium			
3	4.9600		medium			
4	0.4400		small			
5	5.0500		large			
6	0.9600		small			
7	2.8100		medium			
8	4.2900		medium			
9	6.5700		large			
10	3.5400		medium			
11	3.3800		medium			
12	5.7800		large			
13	6.2200		large			
14	4.9400		medium			
15	3.3500		medium			
16						

7. You can save your new data with the Save command from the File menu.

Transform Operators

Transforms use operators to define variables and apply functions. A complete set of arithmetic, relational, and logical operators are provided.

Order of Operation

The order of precedence is consistent with P.E.M.A. (Parentheses, Exponentiation, Multiplication, and Addition) and proceeds as follows, except that parentheses override any other rule:

➤ Exponentiation, associating from right to left

➤ Unary minus

➤ Multiplication and division, associating from left to right

➤ Addition and subtraction, associating from left to right

➤ Relational operators

➤ Logical negation

➤ Logical *and*, associating from left to right

➤ Logical *or*, associating from left to right

This list permits complicated expressions to be written without requiring too many parentheses.

Example: The statement:

a<10 and b<5

groups to (a<10) and (b<5), not to (a<(10 and b))<5.

Σ Note that only parentheses can group terms for processing. Curly and square brackets are reserved for other uses.

Figure 4–1
Examples of Transform
Operators

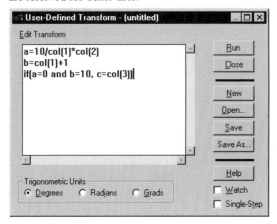

Operations on Ranges

The standard arithmetic operators—addition, subtraction, multiplication, division, and exponentiation—follow basic rules when used with scalars. For operations involving two ranges corresponding entries are added, subtracted, etc., resulting in a range representing the sums, differences, etc., of the two ranges.

If one range is shorter than the other, the operation continues to the length of the longer range, and missing value symbols are used where the shorter range ends.

For operations involving a range and a scalar, the scalar is used against each entry in the range.

Example: The operation:

col(4)*2

produces a range of values, with each entry twice the value of the corresponding value in column 4.

Arithmetic Operators

Arithmetic operators perform arithmetic between a scalar or range and return the result.

+	Add
–	Subtract (also signifies unary minus)
*	Multiply
/	Divide
^ or **	Exponentiate

Multiplication must be explicitly noted with the asterisk. Adjacent parenthetical terms such as (a+b) (c–4) are not automatically multiplied.

Figure 4–2
Arithmetic
Operator Examples

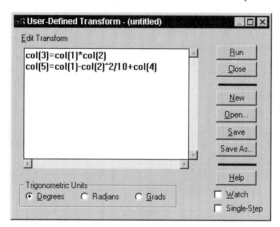

Relational Operators

Relational operators specify the relation between variables and scalars, ranges or equations, or between user-defined functions and equations, establishing definitions, limits and/or conditions.

= or .EQ.	Equal to
> or .GT.	Greater than
>= or .GE.	Greater than or equal to
< or .LT.	Less than
<= or .LE.	Less than or equal to
<>,!=, #, or .NE.	Not equal to

The alphabetic characters can be entered in upper or lower case.

Figure 4–3
Relational and Logical
Operator Examples

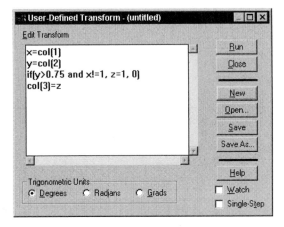

Logical Operators

Logical operators are used to set the conditions for if function statements.

and, &	Intersection
or, |	Union
not, ~	Negation

5

Transform Function Reference

SigmaPlot provides many predefined functions, including arithmetic, statistical, trigonometric, and number-generating functions. In addition, you can define functions of your own.

Function Arguments

Function arguments are placed in parentheses following the function name, separated by commas. Arguments must be typed in the sequence shown for each function.

You must provide the required arguments for each function first, followed by any optional arguments desired. Any omitted optional arguments are set to the default value. Optional arguments are always omitted from right to left. If only one argument is omitted, it will be the last argument. If two are omitted, the last two arguments are set to the default value.

You can use a missing value (i.e., 0/0) as a placeholder to omit an argument.

Example: The col function has three arguments: *column*, *top*, and *bottom*. Therefore, the syntax for the col function is:

col(*column,top,bottom*)

The column number argument is required, but the first (top) and last (bottom) rows are optional, defaulting to row 1 as the first row and the last row with data for the last row.

col(2) returns the entirety of column 2.
col(2,5) returns column 2 from row 5 to the end of the column.
col(2,5,100) returns column 2 from row 5 to row 100.
col(2,0/0,50) returns column 2 from row 1 to the 50th row in the column.

Transform Function Descriptions

The following list groups transforms by function type. It is followed by an alphabetical reference containing complete descriptions of all transform functions and their syntax, with examples.

Worksheet Functions

These worksheet functions are used to specify cells and columns from the worksheet, either to read data from the worksheet for transformation, or to specify a destination for transform results.

Function	Description
block	The block function returns a specified block of cells from the worksheet.
blockheight, blockwidth	The blockheight and blockwidth functions return a specified block of cells or block dimension from the worksheet.
cell	The cell function returns a specific cell from the worksheet.
col	The col function returns a worksheet column or portion of a column.
put into	The put into function places variable or equation results in a worksheet column.
subblock	The subblock function returns a specified block of cells from within another block.

Data Manipulation Functions

The data manipulation functions are used to generate non-random data, and to sample, select, and sort data.

Function	Description
data	The data function generates serial data.
if	The if function conditionally selects between two data sets.
nth	The nth function returns an incremental sampling of data.
sort	The sort function rearranges data in ascending order.

Trigonometric
Functions

SigmaPlot and SigmaStat provide a complete set of trigonometric functions.

Function	Description
arccos, arcsin, arctan	These functions return the arccosine, arcsine, and arctangent of the specified argument.
cos, sin, tan	These functions return the cosine, sine, and tangent of the specified argument.
cosh, sinh, tanh	These functions return the hyperbolic cosine, sine, and tangent of the specified argument.

Numeric Functions

The numeric functions perform a specific type of calculation on a number or range of numbers and returns the appropriate results.

Function	Description
abs	The abs function returns the absolute value.
exp	The exp function returns the values for *e* raised to the specified numbers.
factorial	The factorial function returns the factorial for each specified number.
mod	The mod function returns the modulus, or remainder of division, for specified numerators and divisors.
ln	The ln function returns the natural logarithm for the specified numbers.
log	The log function returns the base 10 logarithm for the specified numbers.
sqrt	The sqrt function returns the square root for the specified numbers.

Range Functions

The following functions give information on ranges.

Function	Description
count	The count function returns the number of numeric values in a range.
missing	The missing function returns the number of missing values and text strings in a range.

Function	Description
size	The size function returns the number of data points in a range, including all numbers, missing values, and text strings.

Accumulation Functions

The accumulation functions return values equal to the accumulated operation of the function.

Function	Description
diff	The diff function returns the differences of the numbers in a range.
sum	The sum function returns the cumulative sum of a range of numbers.
total	The total function returns the value of the total sum of a range.

Random Generation Functions

The two "random" number generating functions can be used to create a series of normally or uniformly distributed numbers.

Function	Description
gaussian	The gaussian function is used to generate a series of *normally* (Gaussian or "bell" shaped) distributed numbers with a specified mean and standard deviation.
random	The random function is used to generate a series of *uniformly* distributed numbers within a specified range.

Precision Functions

The precision functions are used to convert numbers to whole numbers or to round off numbers.

Function	Description
int	The int function converts numbers to integers.
prec	The prec function rounds numbers off to a specified number of significant digits.
round	The round function rounds numbers off to a specified number of decimal places.

Statistical Functions

The statistical functions perform statistical calculations on a range or ranges of numbers.

Function	Description
avg	The avg function calculates the averages of corresponding numbers across ranges. It can be used to calculate the average across rows for worksheet columns.
max, min	The max function returns the largest value in a range; the min function returns the smallest value.
mean	The mean function calculates the mean of a range.
runavg	The runavg function produces a range of running averages.
stddev	The stddev function returns the standard deviation of a range.
stderr	The stderr function calculates the standard error of a range.

Area and Distance Functions

These functions can be used to calculate the areas and distances specified by X,Y coordinates. Units are based on the units used for X and Y.

Function	Description
area	The area function finds the area of a polygon described in X,Y coordinates.
distance	The distance function calculates the distance of a line whose segments are described in X,Y coordinates.
partdist	The partdist function calculates the distances from an initial X,Y coordinate to successive X,Y coordinates in a cumulative fashion.

Curve Fitting Functions

These functions are designed to be used in conjunction with SigmaPlot's nonlinear curve fitter, to allow automatic determination of initial equation parameter estimates from the source data.

You can use these functions to develop your own parameter determination function by using the functions provided with the Standard Regression Equations library provided with SigmaPlot.

Function	Description
ape	This function is used for the polynomials, rational polynomials and other functions which can be expressed as linear functions of the parameters. A linear least squares estimation procedure is used to obtain the parameter estimates.
dsinp	This function returns an estimate of the phase in radians of damped sine functions.
fwhm	This function returns the x width of a peak at half the peak's maximum value for peak shaped functions.
inv	The inv function generates the inverse matrix of an invertible square matrix provided as a block.
lowess	The lowess algorithm is used to smooth noisy data. "Lowess" means *locally weighted regression*. Each point along the smooth curve is obtained from a regression of data points close to the curve point with the closest points more heavily weighted.
lowpass	The lowpass function returns smoothed y values from ranges of x and y variables, using an optional user-defined smoothing factor that uses FFT and IFFT.
sinp	This function returns an estimate of the phase in radians of sinusoidal functions.
x25	This function returns the x value for the y value 25% of the distance from the minimum to the maximum of smoothed data for sigmoidal shaped functions.
x50	This function returns the x value for the y value 50% of the distance from the minimum to the maximum of smoothed data for sigmoidal shaped functions.
x75	This function returns the x value for the y value 75% of the distance from the minimum to the maximum of smoothed data for sigmoidal shaped functions.
xatymax	This function returns the x value for the maximum y in the range of y coordinates for peak shaped functions.
xwtr	This function returns x75-x25 for sigmoidal shaped functions.

Miscellaneous Functions	These functions are specialized functions which perform a variety of operations.

Function	Description
choose	The choose function is the mathematical "n choose r" function.
histogram	The histogram function generates a histogram from a range or column of data.
interpolate	The interpolate function performs linear interpolation between X,Y coordinates.
polynomial	The polynomial function returns results for specified independent variables for a specified polynomial equation.
rgbcolor	The rgbcolor(r,g,b) color function takes arguments r,g, and b between 0 and 255 and returns color to cells in the worksheet.

Special Constructs	Transform constructs are special structures that allow more complex procedures than functions.

Function	Description
for	The for statement is a looping construct used for iterative processing.
if...then...else	The if...then...else construct proceeds along one of two possible series of procedures based on the results of a specified condition.

Fast Fourier Transform Functions	These functions are used to remove noise from and smooth data using frequency-based filtering.

Function	Description
fft	The fft function finds the frequency domain representation of your data.
invfft	The invfft function takes the inverse fft of the data produced by the fft to restore the data to its new filtered form.
real	The real function strips the real numbers out of a range of complex numbers.
img	The img function strips the imaginary numbers out of a range of complex numbers.

Function	Description
complex	The complex function converts a block of real and/or imaginary numbers into a range of complex numbers.
mulcpx	The mulcpx function multiplies two ranges of complex numbers together.
invcpx	The invcpx takes the reciprocal of a range of complex numbers.

abs

Summary   The abs function returns the absolute value for each number in the specified range.

Syntax   abs(*numbers*)

The *numbers* argument can be a scalar or range of numbers. Any missing value or text string contained within a range is ignored and returned as the string or missing value.

Example   The operation col(2) = abs(col(1)) places the absolute values of the data in column 1 in column 2.

ape

Summary   The ape function is used for the polynomials, rational polynomials and other functions which can be expressed as linear functions of the parameters. A linear least squares estimation procedure is used to obtain the parameter estimates. The ape function is used to automatically generate the initial parameter estimates for SigmaPlot's nonlinear curve fitter from the equation provided.

Syntax   ape(*x range,y range,n,m,s,f*)

The *x range* and *y range* arguments specify the independent and dependent variables, or functions of them (e.g., ln(x)). Any missing value or text string contained within one of the ranges is ignored and will not be treated as a data point. *x range* and *y range* must be the same size

The *n* argument specifies the order of the numerator of the equation. The *m* argument specifies the order of the denominator of the equation. n and m must be greater than or equal to 0 ($n, m \geq 0$). If *m* is greater than 0 then *n* must be less than or equal to *m* (if $m > 0$, $n \leq m$).

The s argument specifies whether or not a constant is used. $s=0$ specifies no constant term y_0 in the numerator, $s=1$ specifies a constant term y_0 in the numerator. s must be either 0 or 1. If $n = 0$, s cannot be 0 (there must be a constant).

The number of valid data points must be greater than or equal to $n + m + s$.

The optional f argument defines the amount of Lowess smoothing, and corresponds to the fraction of data points used for each regression. f must be greater than or equal to 0 and less than or equal to 1. $0 \le f \le 1$. If f is omitted, no smoothing is used.

Example   For $x = \{0,1,2\}$, $y=\{0,1,4\}$, the operation col(1)=ape(x,y,1,1,1,0.5]) places the 3 parameter estimates for the equation

$$f(x) = \frac{a + bx}{1 + cx}$$

as the values {5.32907052e-15, 0.66666667, -0.33333333} in column 1.

arccos

Summary   This function returns the inverse of the corresponding trigonometric function.

Syntax   arccos(*numbers*)

The *numbers* argument can be a scalar or range. You can also use the abbreviated function name acos.

The values for the numbers argument must be within -1 and 1, inclusive. Results are returned in degrees, radians, or grads, depending on the Trigonometric Units selected in the User-Defined Transform dialog box. Any missing value or text string contained within a range is ignored and returned as the string or missing value.

The function domain (in radians) is

arccos   0 to π

Example   The operation col(2) = acos(col(1)) places the arccosine of all column 1 data points in column 2.

Related Functions   cos, sin, tan
arcsin, arctan

arcsin

Summary   This function returns the inverse of the corresponding trigonometric function.

Syntax   arcsin(*numbers*)

The *numbers* argument can be a scalar or range. You can also use the abbreviated function name asin.

The values for the numbers argument must be within -1 and 1, inclusive. Results are returned in degrees, radians, or grads, depending on the Trigonometric Units selected in the User-Defined Transform dialog box. Any missing value or text string contained within a range is ignored and returned as the string or missing value.

The function domain (in radians) is:

$$arcsin \quad -\frac{\pi}{2} \; to \; \frac{\pi}{2}$$

Example   The operation col(2) = asin(col(1)) places the arcsine of all column 1 data points in column 2.

Related Functions   cos, sin, tan
arccos, arctan

arctan

Summary   This function returns the inverse of the corresponding trigonometric function.

Syntax   arctan(*numbers*)

The *numbers* argument can be a scalar or range. You can also use the abbreviated function name atan.

The numbers argument can be any value. Results are returned in degrees, radians, or grads, depending on the Trigonometric Units selected in the User-Defined Transform dialog box.

The function domain (in radians) is:

$$arctan \quad -\frac{\pi}{2} \; to \; \frac{\pi}{2}$$

Example   The operation col(2) = atan(col(1)) places the arctangent of all column 1 data points in column 2.

Related Functions   cos, sin, tan
arccos, arcsin

area

Summary   The area function returns the area of a simple polygon. The outline of the polygon is formed by the xy pairs specified in an *x* range and a *y* range.

The list of points does not need to be closed. If the last xy pair does not equal the first xy pair, the polygon is closed from the last xy pair to the first.

The area function only works with simple non-overlapping polygons. If line segments in the polygon cross, the overlapping portion is considered a negative area, and results are unpredictable.

Syntax   area(*x range,y range*)

The *x range* argument contains the x coordinates, and the *y range* argument contains the x coordinates. Corresponding values in these ranges form xy pairs.

If the ranges are uneven in size, excess x or y points are ignored.

Example   For the ranges $x = \{0,1,1,0\}$ and $y = \{0,0,1,1\}$, the operation area (x,y) returns a value of 1. The X and Y coordinates provided describe a square of 1 unit.

Related Functions   dist

avg

Summary   The avg function averages the numbers across corresponding ranges, instead of within ranges. The resulting range is the row-wise average of the range arguments. Unlike the mean function, avg returns a range, not a scalar.

The avg function calculates the arithmetic mean, defined as:

$$\bar{x} = \frac{1}{n} \sum_{i=1}^{n} x_i$$

The avg function can be used to calculate averages of worksheet data across rows rather than within columns.

Syntax   avg($\{x_1,x_2...\},\{y_1,y_2...\},\{z_1,z_2...\}$)

The x_1, y_1, and z_1 are corresponding numbers within ranges. Any missing value or text string contained within a range returns the string or missing value as the result.

Example   The operation avg({1,2,3},{3,4,5}) returns {2,3,4}. 1 from the first range is averaged with 3 from the second range, 2 is averaged with 4, and 3 is averaged with 5. The result is returned as a range.

Related Functions   mean

block

Summary   The block function returns a block of cells from the worksheet, using a range specified by the upper left and lower right cell row and column coordinates.

Syntax   block(*column 1,row 1,column 2,row 2*)

The *column 1* and *row 1* arguments are the coordinates for the upper left cell of the block; the *column 2* and *row 2* arguments are the coordinates for the lower right cell of the block. All values within this range are returned. Operations performed on a block always return a block.

If *column 2* and *row 2* are omitted, then the last row and/or column is assumed to be the last row and column of the data in the worksheet. If you are equating a block to another block, then the last row and/or column is assumed to be the last row and column of the equated block (see the following example).

All column and row arguments must be scalar (not ranges). To use a column title for the column argument, enclose the column title in quotes; block uses the column in the worksheet whose title matches the string.

Example   The command block(5,1) = −block(1,1,3,24) reverses the sign for the values in the range from cell (1,1) to cell (3,24) and places them in a block beginning in cell (5,1).

Related Functions   blockheight, blockwidth
subblock

blockheight, blockwidth

Summary   The blockheight and blockwidth functions return the number of rows or columns, respectively, of a defined block of cells from the worksheet.

Syntax   blockheight(*block*) blockwidth(*block*)

The *block* argument can be a variable defined as a block, or a block function statement.

Example   For the statement x = block(2,1,12,10)

The operation cell(1,1) = blockheight(x) places the number 10 in column 1, row 1 of the worksheet

The operation cell(1,2) = blockwidth(x) places the number 11 in column 1, row 2 of the worksheet.

Related Functions   block
                     subblock

cell

Summary   The cell function returns the contents of a cell in the worksheet, and can specify a cell destination for transform results.

Syntax   cell (*column,row*)

Both *column* and *row* arguments must be scalar (not ranges). To use a column title for the column argument, enclose the column title in quotes; cell uses the column in the worksheet whose title matches the string.

Data placed in a cell inserts or overwrites according to the current insert mode.

Example 1   For the worksheet shown in Figure 5–1, both the operations cell(2,3) and cell("EXP2",3) return a value of 0.5.

Example 2   For the worksheet shown in Figure 5–1, the operation
cell(3,3) = 64^cell(2,3)
raises 64 to the power of the number in cell (2,3), and places the result in cell (3,3).

Related Functions   col

Figure 5–1

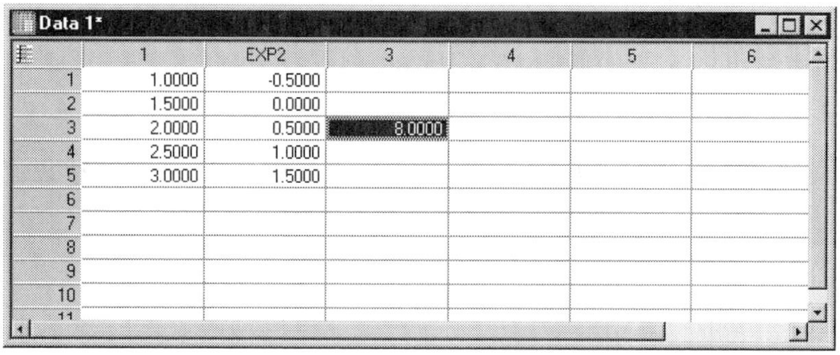

	1	EXP2	3	4	5	6
1	1.0000	-0.5000				
2	1.5000	0.0000				
3	2.0000	0.5000	8.0000			
4	2.5000	1.0000				
5	3.0000	1.5000				
6						
7						
8						
9						
10						
11						

choose

Summary   The choose function determines the number of ways of choosing *r* objects from *n* distinct objects without regard to order.

Syntax   choose(*n*,*r*)

For the arguments *n* and *r*, r < n and "n choose r" is defined as:

$$\binom{n}{r} = \frac{n!}{r!(n-r)!}$$

Examples   To create a function for the binomial distribution, enter the equation:

binomial(p,n,r) = choose(n,r) * (p^r) * (1–p) ^ (n–r)

col

Summary   The col function returns all or a portion of a worksheet column, and can specify a column destination for transform results.

Syntax   col (*column,top,bottom*)

The *column* argument is the column number or title. To use a column title for the column argument, enclose the title in quotation marks. The *top* and *bottom* arguments specify the first and last row numbers, and can be omitted. The default row numbers are 1 and the end of the column, respectively; if both are omitted, the entire column is used. All parameters must be scalar. Data placed in a column inserts or overwrites according to the current insert mode.

Example 1   For the worksheet shown in Figure 5–1, the operation col(3) returns the entire range of five values, the operation col(3,4) returns {8.9, 9.1}, and the operation col("data2",2,3) returns {7.9,8.4}.

Example 2   For the worksheet shown in Figure 5–1, the operation col(4) = col(3)*2 multiples all the values in column 3 and places the results in column 4.

Related Functions cell

Figure 5-2

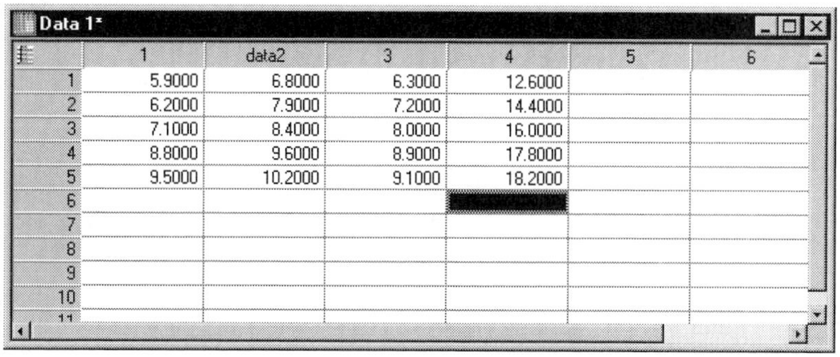

complex

Summary Converts a block of real and imaginary numbers into a range of complex numbers.

Syntax complex (*range,range*)

The first range contains the real values, the second range contains the imaginary values and is optional. If you do not specify the second range, the complex transform returns zeros for the imaginary numbers. If you do specify an imaginary range, it must contain the same number of values as the real value range.

Example If x = {1,2,3,4,5,6,7,8,9,10}, the operation complex(x) returns {{1,2,3,4,....,9,10}, {0,0,0,0,....,0,0}}.

If x = {1.0,-0.75,3.1} and y = {1.2,2.1,-1.1}, the operation complex(x,y) returns {{1.0,-0.75,3.1}, {1.2,2.1,-1.1}}.

Related Functions fft, invfft, real, imaginary, mulcpx, invcpx

cos

Summary This function returns ranges consisting of the cosine of each value in the argument given.

This and other trigonometric functions can take values in radians, degrees, or grads. This is determined by the Trigonometric Units selected in the User-Defined Transform dialog box.

Syntax   cos(*numbers*)

The *numbers* argument can be a scalar or range.

If you regularly use values outside of the usual -2π to 2π (or equivalent) range, use the mod function to prevent loss of precision. Any missing value or text string contained within a range is ignored and returned as the string or missing value.

Example   If you choose Degrees as your Trigonometric Units in the User-Defined Transform dialog box, the operation cos({0,60,90,120,180}) returns values of {1,0.5,0,−0.5,−1}.

Related Functions   acos, asin, atan
sin, tan

cosh

Summary   This function returns the hyperbolic cosine of the specified argument.

Syntax   cosh(*numbers*)

The *numbers* argument can be a scalar or range.

Like the circular trig functions, this function also accepts numbers in degrees, radians, or grads, depending on the units selected in the User-Defined Transform dialog box. Any missing value or text string contained within a range is ignored and returned as the string or missing value.

Example   The operation x = cosh(col(2)) sets the variable x to be the hyperbolic cosine of all data in column 2.

Related Functions   sinh, tanh

count

Summary   The count function returns the value or range of values equal to the number of non-missing numeric values in a range. Missing values and text strings are not counted.

Syntax   count(*range*)

The *range* argument must be a single range (indicated with the { } brackets) or a worksheet column.

Examples   For the worksheet in Figure 5–1:

the operation count(col(1)) returns a value of 5,
the operation count(col(2)) returns a value of 6, and
the operation count(col(3)) returns a value of 0.

Related Functions   missing, size

Figure 5–3

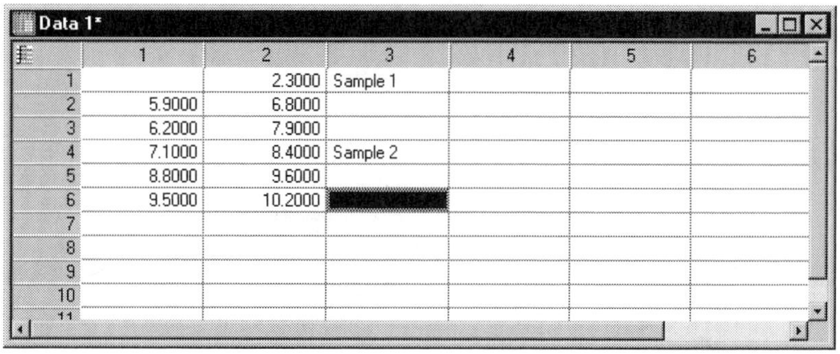

	1	2	3	4	5	6
1		2.3000	Sample 1			
2	5.9000	6.8000				
3	6.2000	7.9000				
4	7.1000	8.4000	Sample 2			
5	8.8000	9.6000				
6	9.5000	10.2000				
7						
8						
9						
10						
11						

Data 1*

data

Summary   The data function generates a range of numbers from a starting number to an end number, in specified increments.

Syntax   data(*start,stop,step*)

All arguments must be scalar. The *start* argument specifies the beginning number and the *end* argument sets the last number.

If the *step* parameter is omitted, it defaults to 1. The start parameter can be more than or less than the stop parameter. In either case, data steps in the correct direction. Remainders are ignored.

Examples   The operation data(1,5) returns the range of values {1,2,3,4,5}.
The operation data(10,1,2) returns the values {10,8,6,4,2}.

Note that if start and stop are equal, this function produces a number of copies of start equal to step. For example, the operation data(1,1,4) returns {1,1,1,1}.

Related Functions   size, [ ] array reference

diff

Summary : The diff function returns a range or ranges of numbers which are the differences between a given number in a range and the preceding number. The value of the preceding number is subtracted from the value of the following number.

Because there is no preceding number for the first number in a range, the value of the first number in the result is always the same as the first number in the argument range.

Syntax : diff(*range*)

The *range* argument must be a single range (indicated with the { } brackets) or a worksheet column. Any missing value or text string contained within the range is returned as the string or missing value.

Examples : For x = {9,16,7}, the operation diff(x) returns a value of {9,7,−9}.
For y = {4,−6,12}, the operation diff(y) returns a value of {4,−10,18}.

Related Functions : sum, total

dist

Summary : The dist function returns a scalar representing the distance along a line. The line is described in segments defined by the X,Y pairs specified in an *x* range and a *y* range.

Syntax : dist(*x range,y range*)

The *x* range argument contains the X coordinates, and the *y* range argument contains the Y coordinates. Corresponding values in these ranges form X,Y pairs. If the ranges are uneven in size, excess X or Y points are ignored.

Example : For the ranges x ={0,1,1,0,0} and y = {0,0,1,1,0}, the operation dist(x,y) returns 4.0. The X and Y coordinates provided describe a square of 1 unit *x* by 1 unit *y*.

Related Functions : partdist

dsinp

Summary : The dsinp function automatically generates the initial parameter estimates for a damped sinusoidal functions using the FFT method. The four parameter estimates are returned as a vector.

Syntax   dsinp(*x range, y range*)

The *x range* argument specifies the x variable, and the y range argument specifies the y variable. Any missing value or text string contained within one of the ranges is ignored and will not be treated as a data point. *x range* and *y range* must be the same size, and the number of valid data points must be greater than or equal to 3.

Σ   dsinp is especially used to estimate parameters on waveform functions. This is only useful when this function is used in conjunction with nonlinear regression.

Related Functions   sinp

exp

Summary   The exp function returns a range of values consisting of the number *e* raised to each number in the specified range. This is numerically identical to the expression e^(*numbers*), but uses a faster algorithm.

Syntax   exp(*numbers*)

The *numbers* argument can be a scalar or range of numbers. Any missing value or text string contained within a range is ignored and returned as the string or missing value.

Example   The operation exp(1) returns a value of 2.718281828459045.

Related Functions   ln

factorial

Summary   The factorial function returns the factorial of a specified range.

Syntax   factorial({*range*})

The *range* argument must be a single range (indicated with the { } brackets) or a worksheet column. Any missing value or text string contained within a range is ignored and returned as the string or missing value. Non-integers are rounded down to the nearest integer or 1, whichever is larger.

For factorial(*x*):

$x < 0$ returns a missing value,
$0 \leq x < 180$ returns x!, and
$x \geq 180$ returns $+\infty$

Example 1   The operation factorial({1,2,3,4,5}) returns {1,2,6,24,120}.

Example 2 To create a transform equation function for the Poisson distribution, you can type:

Poisson(m,x)=(m^x)*exp(−m)/factorial(x)

fft

Summary The fft function finds the frequency domain representation of your data using the Fast Fourier Transform.

Syntax fft(*range*)

The parameter can be a range of real values or a block of complex values. For complex values there are two columns of data. The first column contains the real values and the second column represents the imaginary values. This function works on data sizes of size 2^n numbers. If your data set is not 2^n in length, the fft function pads 0 at the beginning and end of the data range to make the length 2^n.

The fft function returns a range of complex numbers.

Example For x = {1,2,3,4,5,6,7,8,9,10}, the operation fft(x) takes the Fourier transform of the ramp function with real data from 1 to 10 with 3 zeros padded on the front and back and returns a 2 by 16 block of complex numbers.

Related Functions invfft, real, imaginary, complex, mulcpx, invcpx

for

Summary The for statement is a looping construct used for iterative processing.

Syntax for *loop variable* = *initial value* to *end value* step *increment* do
equation
equation
.
.
.
end for

Transform equation statements are evaluated iteratively within the for loop. When a for statement is encountered, all functions within the loop are evaluated separately from the rest of the transform.

The *loop variable* can be any previously undeclared variable name. The *initial value* for the loop is the beginning value to be used in the loop statements. The *end value* for the loop variable specifies the last value to be processed by the for statement. After the end value is processed, the loop is terminated. In addition, you can specify a loop variable step *increment*, which is used to "skip" values when proceeding from the initial value to end value. If no increment is specified, an increment of 1 is assumed.

Σ You must separate *for*, *to*, *step*, *do*, *end for*, and all condition statement operators, variables and values with spaces.

The for loop statement is followed by a series of one or more transform equations which process the loop variable values.

Inside for loops, you can:

➤ indent equations

➤ nest for loops

Note that these conditions are allowed only within for loops. You cannot redefine variable names within for loops.

Example 1 The operation:

for i = 1 to size(col(1)) do
cell(2,i) = cell(1,i)*i
end for

multiplies all the values in column 1 by their row number and places them in column 2.

Example 2 The operation:

for j = cell(1,1) to cell (1,64) step 2 do
col(10) = col(9)^j
end for

Takes the value from cell (1,1) and increments by 2 until the value in cell (1,64) is reached, raises the data in column 9 to that power, and places the results in column 10.

fwhm

Summary The fwhm function returns value of the x width at half-maxima in the ranges of coordinates provided, with optional Lowess smoothing.

Syntax fwhm(*x range, y range, f*)

The *x range* argument specifies the x variable, and the y range argument specifies the y variable. Any missing value or text string contained within one of the ranges is ignored and will not be treated as a data point. *x range* and *y range* must have the same size, and the number of valid data points must be greater than or equal to 3.

The optional *f* argument defines the amount of Lowess smoothing, and corresponds to the fraction of data points used for each regression. *f* must be greater than or equal to 0 and less than or equal to 1. $0 \le f \le 1$. If *f* is omitted, no smoothing is used.

Example   For x = {0,1,2}, y={0,1,4}, the operation

col(1)=fwhm(x,y)

places the x width at half-maxima 1.00 into column 1.

Related Functions   xatymax

gaussian

Summary   This function generates a specified number of normally (Gaussian or "bell" shaped) distributed numbers from a seed number, using a supplied mean and standard deviation.

Syntax   gaussian(*number,seed,mean,stddev*)

The *number* argument specifies how many random numbers to generate.

The *seed* argument is the random number generation seed to be used by the function. If you want to generate a different random number sequence each time the function is used, enter 0/0 for the seed. Enter the same number to generate an identical random number sequence. If the seed argument is omitted, a randomly selected seed is used.

The *mean* and *stddev* arguments are the mean and standard deviation of the normal distribution curve, respectively. If mean and stddev are omitted, they default to 0 and 1.

Note that function arguments are omitted from right to left. If you want to specify a stddev, you must either specify the mean argument or omit it by using 0/0.

Example   The operation gaussian(100) uses a seed of 0 to produce 100 normally distributed random numbers, with a mean of 0.0 and a standard deviation of 1.0.

Related Functions   random

histogram

Summary   The histogram function produces a histogram of the values range in a specified range, using a defined interval set.

Syntax histogram(*range, buckets*)

The *range* argument must be a single range (indicated with the { } brackets) or a worksheet column. Any missing value or text string contained within a range is ignored.

The *buckets* argument is used to specify either the number of evenly incremented histogram intervals, or both the number and ranges of the intervals. This value can be scalar or a range. In both versions, missing values and strings are ignored.

If the buckets parameter is a scalar, it must be a positive integer. A scalar buckets argument generates a number of intervals equal to the buckets value. The histogram intervals are evenly sized; the range is the minimum value to the maximum value of the specified range.

If the buckets argument is specified as a range, each number in the range becomes the upper bound (inclusive) of an interval. Values from $-\infty$ to $\leq$ the first bucket fall in the first histogram interval, values from > first bucket to $\leq$ second bucket fall in the second interval, *etc*. The buckets range must be strictly increasing in value. An additional interval is defined to catch any value which does not fall into the defined ranges. The number of values occurring in this extra interval (including 0, or no values outside the range) becomes the last entry of the range produced by histogram function.

Example 1 For col(1) = {1,20,30,35,40,50,60}, the operation col(2) = histogram(col(1),3) places the range {2,3,2} in column 2. The bucket intervals are automatically set to 20, 40, and 60, so that two of the values in column 1 fall under 20, three fall under 40, and two fall under 60.

Example 2 For buckets = {25,50,75}, the operation col(3) = histogram(col(1),buckets) places {2,4,1,0} in col(3). Two of the values in column 1 fall under 25, four fall under 50, one under 75, and no values fall outside the range.

if

Summary The if function either selects one of two values based on a specified condition, or proceeds along a series of calculations bases on a specified condition.

Syntax if(*condition, true value, false value*)

The *true value* and *false value* arguments can be any scalar or range. For a true *condition*, the true value is returned; for a false condition, the false value is returned. If the false value argument is omitted, a false condition returns a missing value.

If the condition argument is scalar, then the entire true value or false value argument is returned. If the condition argument contains a range, the result is a new range. For each true entry in the condition range, the corresponding entry in the true value argument is returned. For a false entry in the condition range, the corresponding entry in false value is returned.

If the false value is omitted and the condition entry is false, the corresponding entry in the true value range is omitted. This can be used to conditionally extract data from a range.

Example 1   The operation col(2) = if(col(1)< 75,"FAIL","PASS") reads in the values from column 1, and places the word "FAIL" in column 2 if the column 1 value is less than 75, and the word "PASS" if the value is 75 or greater.

Example 2   For the operation y = if(x < 2 or x > 4,99,x), an *x* value less than 2 or greater than 4 returns a *y* value of 99, and all other *x* values return a *y* value equal to the corresponding *x* value.

If you set x = {1,2,3,4,5}, then *y* is returned as {99,2,3,4,99}. The condition was true for the first and last *x* range entries, so 99 was returned. The condition was false for x = 2, 3, and 4, so the *x* value was returned for the second, third, and fourth *x* values.

if...then...else

Summary   The if...then...else function proceeds along one of two possible series of calculations based on a specified condition.

Syntax   if *condition* then
    statement
    statement...
else
    statement
    statement...
end if

To use the if...then...else construct, follow the if *condition* then statement by one or more transform equation statements, then specify the else statement(s). When an if...then...else statement is encountered, all functions within the statement are evaluated separately from the rest of the transform.

Σ You must separate if, then, and all condition statement operators, variables, and values with spaces.

Inside if...then...else constructs, you can:

➤   type more than one equation on a line

➤ indent equations

➤ nest additional if constructs

Note that these conditions are allowed only within if...else statements. You cannot redefine variable names within an if...then...else construct.

Example   The operations:

i = cell(1,1)
j = cell(1,2)
If i < 1 and j > 1 then x = col(3)
else x = col(4)
end if

sets x equal to column 3 if i is less than 1 and j is greater than 1; otherwise, x is equal to column 4.

imaginary (img)

Summary   The imaginary function strips the imaginary values out of a range of complex numbers.

Syntax   img(*block*)

The range is made up of complex numbers.

Example   If x = {{1,2,3,4,5,6,7,8,9,10}, {0,0,0,....0,0}}, the operation img(x) returns {0,0,0,0,0,0,0,0,0,0}.

If x = {{1.0,-0.75, 3.1}, {1.2,2.1,-1.1}}, the operation img(x) returns {1.2,2.1,-1.1}.

Related Functions   real, fft, invfft, complex, mulcpx, invcpx

int

Summary   The int function returns a number or range of numbers equal to the largest integer less than or equal to each corresponding number in the specified range. All numbers are rounded down to the nearest integer.

Syntax   int(*numbers*)

The numbers argument can be a scalar or range of numbers. Any missing value or text string contained within a range is ignored and returned as the string or missing value.

Example The operation int({.9,1.2,2.2,−3.8}) returns a range of {0.0,1.0,2.0,−4.0}.

Related Functions prec, round

interpolate

Summary The interpolate function performs linear interpolation on a set of X,Y pairs defined by an *x* range and a *y* range. The function returns a range of interpolated *y* values from a range of values between the minimum and maximum of the *x* range.

Syntax interpolate(*x range,y range,range*)

Values in the *x range* argument must be strictly increasing or strictly decreasing.

The *range* argument must be a single range (indicated with the {} brackets) or a worksheet column. Missing values and text strings are not allowed in the *x range* and *y range*. Text strings in *range* are replaced by missing values.

Extrapolation is not possible; missing value symbols are returned for range argument values less than the lowest *x* range value or greater than the highest *x* range value.

Examples For x = {0,1,2}, y = {0,1,4}, and range = data(0,2,.5) (this data operation returns numbers from 0 to 2 at increments of 0.5), the operation
col(1) = interpolate(x,y,range) places the range {0.0,0.5,1.0,2.5,4.0}
into column 1.

If *range* had included values outside the range for *x*, missing values would have been returned for those out-of-range values.

inv

Summary The inv function generates the inverse matrix of an invertible square matrix provided as a block.

Syntax inv(*block*)

The *block* argument is a block of numbers with real values in the form of a square matrix. The number of rows must equal the number of columns.

The function returns a block of numbers with real values in the form of the inverse of the square matrix provided.

Example   For the matrix:

1.00	3.00	4.00
2.00	1.00	3.00
3.00	4.00	2.00

in block(2,3,4,5) the operation block(2,7)=inv(block(2,3,4,5)) generates the inverse matrix:

-0.40	0.40	0.20
0.20	-0.40	0.20
0.20	0.20	-0.20

in block (2,7,4,9).

invcpx

Summary   This function takes the reciprocal of a range of complex numbers.

Syntax   invcp(*block*)

The input and output are blocks of complex numbers. The invcpx function returns the range 1/c for each complex number in the input block.

Example   If x = complex ({3,0,1}, {0,1,1}), the operation invcpx(x) returns {{0.33333, 0.0, 0.5}, {0.0,−1.0,−0.5}}.

Related Functions   fft, invfft, real, imaginary, complex, mulcpx

invfft

Summary   The inverse fft function (invfft) takes the inverse Fast Fourier Transform (fft) of the data produced by the fft to restore the data to its new filtered form.

Syntax   invfft(*block*)

The parameter is a complex block of spectral numbers with the real values in the first column and the imaginary values in the second column. This data is usually generated from the fft function. The invfft function works on data sizes of size 2^n numbers. If your data set is not 2^n in length, the invfft function pads 0 at the beginning and end of the data range to make the length 2^n.

The function returns a complex block of numbers.

Example   If $x = \{\{1,2,3,...,9,10\}, \{0,0,0,...,0,0\}\}$, the operation invfft(fft(x)) returns $\{\{0,0,0,1,2,3,...,9,10,0,0,0\}, \{0,0,0,...0,0\}$.

Related Functions   fft, real, imaginary, complex, mulcpx, invcpx

ln

Summary   The ln function returns a value or range of values consisting of the natural logarithm of each number in the specified range.

Syntax   ln(*numbers*)

The *numbers* argument can be a scalar or range of numbers. Any missing value or text string contained within a range is ignored and returned as the string or missing value.

For $\ln(x)$:

$x < 0$ returns an error message, and
$x = 0$ returns $-\infty$

The largest value allowed is approximately $x < 10^{4933}$.

Example   The operation ln(2.71828) returns a value ≈ 1.0.

Related Functions   exp

log

Summary   The log function returns a value or range of values consisting of the base 10 logarithm of each number in the specified range.

Syntax   log(*numbers*)

The *numbers* argument can be a scalar or range of numbers. Any missing value or text string contained within a range is ignored and returned as the string or missing value.

For $\log(x)$:

$x < 0$ returns an error message,
$x = 0$ returns $-\infty$

The largest value allowed is approximately $x < 10^{4933}$.

Example   The operation log(100) returns a value of 2.

lookup

Summary   The lookup function compares values with a specified table of boundaries and returns either a corresponding index from a one-dimensional table, or a corresponding value from a two-dimensional table.

Syntax   lookup(*numbers,x table,y table*)

The *numbers* argument is the range of values looked up in the specified *x table*. The x table argument consists of the upper bounds (inclusive) of the *x* intervals within the table and must be ascending in value. The lower bounds are the values of the previous numbers in the table ($-\infty$ for the first interval).

You must specify numbers and an x table. If only the numbers and x table arguments are specified, the lookup function returns an index number corresponding to the x table interval; the interval from $-\infty$ to the first boundary corresponds to an index of 1, the second to 2, *etc.*

If a number value is larger than the last entry in x table, lookup will return a missing value as the index. You can avoid missing value results by specifying 1/0 (infinity) as the last value in x table.

The optional *y table* argument is used to assign *y* values to the *x* index numbers. The y table argument must be the same size as the x table argument, but the elements do not need to be in any particular order. If y table is specified, lookup returns the y table value corresponding to the x table index value, *i.e.,* the first y table value for an index of 1, the second y table value for an index of 2, *etc.*

Σ Note that the *x table* and *y table* ranges correspond to what is normally called a "lookup table."

Example 1   For n={-4,11,31} and x={1,10,30}, **col(1)=lookup(n,x)** places the index values of 1, 3, and -- (missing value) in column1.

index #	1	2	3
x table	1	10	30

-4 ───────────┐

11 ──────────────────────────────┘

31 ──── (missing value)

-4 falls beneath 1, or the first x boundary; 11 falls beyond 10 but below 30, and 31 lies beyond 30.

Example 2   To generate triplet values for the range {9,6,5}, you can use the expression lookup(data(1/3,3,1/3),data(1,3),{9,6,5}) to return {9,9,9,6,6,6,5,5,5}. This looks up the numbers 1/3, 2/3, 1, 1 1/3, 1 2/3, 2, 2 1/3, 22/3, and 3 using x table boundaries 1, 2, and 3 and corresponding y table values 9, 6, and 5.

y table	9	6	5
x table	1	2	3

1/3 ─────────┐

2/3 ─────────┤

1 ──────────┘

1 1/3 ──────────────────┐

1 2/3 ──────────────────┤

2 ──────────────────────┘

2 1/3 ──────────────────────────────┐

22/3 ───────────────────────────────┤

3 ─────────────────────────────────┘

lowess

Summary
: The lowess function returns smoothed y values as a range from the ranges of x and y variables provided, using a user-defined smoothing factor. "Lowess" means *locally weighted regression*. Each point along the smooth curve is obtained from a regression of data points close to the curve point with the closest points more heavily weighted.

Syntax
: lowess(*x range, y range, f*)

  The *x range* argument specifies the x variable, and the *y range* argument specifies the y variable. Any missing value or text string contained within one of the ranges is ignored and will not be treated as a data point. *x range* and *y range* must be the same size, and the number of valid data points must be greater than or equal to 3.

  The *f* argument defines the amount of Lowess smoothing, and corresponds to the fraction of data points used for each regression. *f* must be greater than or equal to 0 and less than or equal to 1. $0 \leq f \leq 1$. Note that unlike lowpass, lowess requires an *f* argument.

Example
: For x = {1,2,3,4}, y={0.13, 0.17, 0.50, 0.60}, the operation

  col(1)=lowess(x,y,1)

  places the smoothed y data 0.10, 0.25, 0.43, 0.63 into column 1.

Related Functions
: lowpass

lowpass

Summary
: The lowpass function returns smoothed y values from ranges of x and y variables, using an optional user-defined smoothing factor that uses FFT and IFFT.

Syntax
: lowpass(*x range, y range, f*)

  The *x range* argument specifies the x variable, and the *y range* argument specifies the y variable. Any missing value or text string contained within one of the ranges is ignored and will not be treated as a data point. *x range* and *y range* must be the same size, and the number of valid data points must be greater than or equal to 3.

  The optional *f* argument defines whether FFT and IFFT are used. *f* must be greater than or equal to 0 and less than or equal to 100 ($0 \leq f \leq 100$). If *f* is omitted, no Fourier transformation is used.

Σ lowpass is especially designed to perform smoothing on waveform functions as a part of nonlinear regression.

Example For x = {0,1,2}, y={0,1,4}, the operation

col(1)=lowpass(x,y,88)

places the newly smoothed data 0.25, 1.50, 2.25 into column 1.

Related Functions lowess

max

Summary The max function returns the largest number found in the range specified.

Syntax max(*range*)

The *range* argument must be a single range (indicated with the { } brackets) or a worksheet column. Any missing value or text string contained within a range is ignored.

Example For x = {7,4,−4,5}, the operation max(x) returns a value of 7, and the operation min(x) returns a value of −4.

mean

Summary The mean function returns the average of the range specified. Use this function to calculate column averages (as opposed to using the avg function to calculate row averages).

The mean function calculates the arithmetic mean, defined as:

$$x = \frac{1}{n} \sum_{i=1}^{n} x_i$$

Syntax mean(*range*)

The *range* argument must be a single range (indicated with the {} brackets) or a worksheet column. Any missing value or text string contained within a range is ignored.

Example The operation mean({1,2,3,4}) returns a value of 2.5.

Related Functions avg

min

Summary   The min function returns the smallest number in the range specified.

Syntax   min(*range*)

The *range* argument must be a single range (indicated with the {} brackets) or a worksheet column. Any missing value or text string contained within a range is ignored.

Example   For x = {7,4,−4,5}, the operation max(x) returns a value of 7, and the operation min(x) returns a value of −4.

missing

Summary   The missing function returns a value or range of values equal to the number of missing values and text strings in the specified range.

Syntax   missing(*range*)

The *range* argument must be a single range (indicated with the {} brackets) or a worksheet column.

Example   For Figure 5–1, the operation missing(col(1)) returns a value of 1, the operation missing(col(2)) returns a value of 0, and the operation missing(col(3)) returns a value of 4.

Related Functions   count, size

Figure 5–4

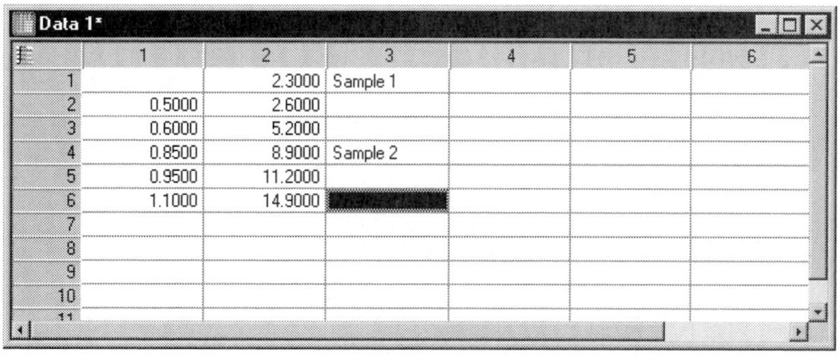

	1	2	3	4	5	6
1		2.3000	Sample 1			
2	0.5000	2.6000				
3	0.6000	5.2000				
4	0.8500	8.9000	Sample 2			
5	0.9500	11.2000				
6	1.1000	14.9000				
7						
8						
9						
10						
11						

mod

Summary   The mod function returns the modulus (the remainder from division) for corresponding numbers in numerator and divisor arguments.

This is the real (not integral) modulus, so both ranges may be nonintegral values.

Syntax   mod(*numerator,divisor*)

The *numerator* and *divisor* arguments can be scalars or ranges. Any missing value or text string contained within a range is returned as the string or missing value.

For any divisor ≠ 0, the mod function returns the remainder of $\frac{numerator}{divisor}$.

For mod(x,0), that is, for ***divisor*** = 0,

x > 0 returns +∞
x = 0 returns +∞
x < 0 returns −∞

Example   The operation mod({4,5,4,5},{2,2,3,3}) returns the range {0,1,1,2}. These are the remainders for 4÷2, 5÷2, 4÷3, and 5÷3.

mulcpx

Summary   The mulcpx function multiplies two blocks of complex numbers together.

Syntax   mulcpx(*block, block*)

Both input blocks should be the same length. The mulcpx function returns a block that contains the complex multiplication of the two ranges.

Example   If u = {{1,1,0},{0,1,1}}, the operation mulcpx(u,u) returns {{1,0,−1}, {0,2,0}}.

Related Functions   fft, invfft, real, imaginary, complex, invcpx

nth

Summary   The nth function returns a sampling of a provided range, with the frequency indicated by a scalar number. The result always begins with the first entry in the specified range.

Syntax   nth(*range,increment*)

The *range* argument is either a specified range (indicated with the { } brackets) or a worksheet column. The *increment* argument must be a positive integer.

Example   The operation **col(1)=nth({1,2,3,4,5,6,7,8,9,10},3)** places the range { 1,4,7,10} in column 1. Every third value of the range is returned, beginning with 1.

partdist

Summary   The partdist function returns a range representing the distance from the first X,Y pair to each other successive pair. The line segment X,Y pairs are specified by an *x* range and a *y* range.

The last value in this range is numerically the same as that returned by dist, assuming the same *x* and *y* ranges.

Syntax   partdist*(x range,y range)*

The *x range* argument specifies the x coordinates, and the *y range* argument specifies the y coordinates. Corresponding values in these ranges form xy pairs.

If the ranges are uneven in size, excess x or y points are ignored.

Example   For the ranges x = {0,1,1,0,0} and y = {0,0,1,1,0}, the operation **partdist(x,y)** returns a range of {0,1,2,3,4}. The X and Y coordinates provided describe a square of 1 unit *x* by 1 unit *y*.

Related Functions   dist

polynomial

Summary   The polynomial function returns the results for independent variable values in polynomials. Given the coefficients, this function produces a range of *y* values for the corresponding *x* values in range.

The function takes one of two forms. The first form has two arguments, both of which are ranges. Values in the first range are the independent variable values. The second range represents the coefficients of the polynomial, with the constant coefficient listed first, and the highest order coefficient listed last.

The second form accepts two or more arguments. The first argument is a range consisting of the independent variable values. All successive arguments are scalar and represent the coefficients of a polynomial, with the constant coefficient listed first and the highest order coefficient listed last.

Syntax   polynomial(*range,coefficents*) or
          polynomial(*range,a0,a1,...,an*)

The *range* argument must be a single range (indicated with the {} brackets) or a worksheet column. Text strings contained within a range are returned as a missing value.

The *coefficients* argument is a range consisting of the polynomial coefficient values, from lowest to highest. Alternately, the coefficients can be listed individually as scalars.

Example   To solve the polynomial $y = x^2 + x + 1$ for x values of 0, 1, and 2, type the equation **polynomial({0,1,2},1,1,1)**. Alternately, you could set x ={1,1,1}, then enter **polynomial({0,1,2},x)**. Both operations return a range of {1,3,7}.

prec

Summary   The prec function rounds a number or range of numbers to the specified number of significant digits, or places of significance. Values are rounded to the nearest integer; values of exactly 0.5 are rounded up.

Syntax   prec(*numbers,digits*)

The *numbers* argument can be a scalar or range of numbers. Any missing value or text string contained within a range is ignored and returned as the string or missing value.

If the *digits* argument is a scalar, all numbers in the range have the same number of places of significance. If the digits argument is a range, the number of places of significance vary according to the corresponding range values. If the size of the digits range is smaller than the numbers range, the function returns missing values for all numbers with no corresponding digits.

Example   For x = {13570,3.141,.0155,999,1.92}, the operation **prec(x,2)** returns {14000,3.100,.0160,1000,1.90}.

For y = {123.5,123.5,123.5,123.5}, the operation **prec(y,{1,2,3,4})** returns {100.0, 120.0,124.0,123.5}.

Related Functions   int, round

put into

Summary   The put into function places calculation results in a designated column on the worksheet. It operates faster than the equivalent equality relationship.

Syntax put *results* into col*(column)*

The *results* argument can be either the result of an equation, function or variable. The *column* argument is either the column number of the destination column, or the column title, enclosed in quotes.

Data put into columns inserts or overwrites according to the current insert mode.

Example To place the results of the equation y = data(1,100) in column 1, you can type **col(1) = y**. However, entering **put y into col(1)** runs faster.

Related Functions col = (arithmetic) operator

random

Summary This function generates a specified number of uniformly distributed numbers within the range.

Rand and rnd are synonyms for the random function.

Syntax random*(number,seed,low,high)*

The *number* argument specifies how many random numbers to generate.

The *seed* argument is the random number generation seed to be used by the function. If you want to generate a different random number sequence each time the function is used, enter 0/0 for the seed. If the seed argument is omitted, a randomly selected seed is used.

The *low* and *high* arguments specify the beginning and end of the random number distribution range. The low boundary is included in the range. If low and high are omitted, they default to 0 and 1, respectively.

Note that function arguments are omitted from right to left. If you want to specify a high boundary, you must specify the low boundary argument first.

Example The operation **random(50,0/0,1,7)** produces 50 uniformly distributed random numbers between 1 and 7. The sequence is different each time this random function is used.

Related Functions gaussian

real

Summary The real function strips the real values from a complex block of numbers.

Syntax real (*range*)

The *range* argument consists of complex numbers.

Example If x = complex ({1,2,3,...,9,10}, {0,0,...,0}), the operation real(x) returns {1,2,3,4,5,6,7,8,9,10}, leaving the imaginary values out.

Related Functions fft, invfft, imaginary, complex, mulcpx, invcpx

rgbcolor

Summary The transform function rgbcolor takes arguments r, g, and b between 0 and 255 and returns the corresponding color to cells in the worksheet. This function can be used to apply custom colors to any element of a graph or plot that can use colors chosen from a worksheet column.

Syntax rgbcolor*(r,g,b)*

The *r,g,b* arguments define the red, green, and blue intensity portions of the color. These values must be scalars between 0 and 255. Numbers for the arguments less than 0 or greater than 255 are truncated to these values.

Examples The operation **rgbcolor(255,0,0)** returns red.
The operation **rgbcolor(0,255,0)** returns green.
The operation **rgbcolor(0,0,255)** returns blue.

The following statements place the secondary colors yellow, magenta, and cyan into rows 1, 2, and 3 into column 1:

cell(1,1)=rgbcolor(255,255,0)
cell(1,2)=rgbcolor(255,0,255)
cell(1,3)=rgbcolor(0,255,255)

Shades of gray are generated using equal arguments. To place black, gray, and white in the first three rows of column 1:

cell(1,1)=rgbcolor(0,0,0)
cell(1,2)=rgbcolor(127,127,127)
cell(1,3)=rgbcolor(255,255,255)

round

Summary   The round function rounds a number or range of numbers to the specified decimal places of accuracy. Values are rounded up or down to the nearest integer; values of exactly 0.5 are rounded up.

Syntax   round*(numbers,places)*

The *numbers* argument can be a scalar or range of numbers. Any missing value or text string contained within a range is ignored and returned as the string or missing value.

If the *places* argument is negative, rounding occurs to the left of the decimal point. To round to the nearest whole number, use a places argument of 0.

Examples   The operation **round(92.1541,2)** returns a value of 92.15.
The operation **round(0.19112,1)** returns a value of 0.2.
The operation **round(92.1541,–2)** returns a value of 100.0.

Related Functions   int, prec

runavg

Summary   The runavg function produces a range of running averages, using a window of a specified size as the size of the range to be averaged. The resulting range is the same length as the argument range.

Syntax   runavg*(range,window)*

The *range* argument must be a single range (indicated with the {} brackets) or a worksheet column. Any missing value or text string contained within a range is replaced with 0.

If the *window* argument is even, the next highest odd number is used. The tails of the running average are computed by appending

$\frac{(window - 1)}{2}$ additional initial and final values to their respective ends of range.

Example   The operation runavg({1,2,3,4,5},3) returns {1.33,2,3,4,4.67}.

The value of the window argument is 3, so the first result value is calculated as:

$$\frac{\dfrac{(3-1)}{2} + 1 + 2}{3}$$

The second value is calculated as:

$$\frac{1 + 2 + 3}{3} \quad \text{, etc.}$$

Related Functions   avg

mean

sin

Summary   This function returns ranges consisting of the sine of each value in the argument given.

This and other trigonometric functions can take values in radians, degrees, or grads. This is determined by the Trigonometric Units selected in the User-Defined Transform dialog box.

Syntax   sin(***numbers***)

The ***numbers*** argument can be a scalar or range.

If you regularly use values outside of the usual -2π to 2π (or equivalent) range, use the **mod** function to prevent loss of precision. Any missing value or text string contained within a range is ignored and returned as the string or missing value.

Example   If you choose Degrees as your Trigonometric Units in the transform dialog box, the operation **sin({0,30,90,180,270})** returns values of {0,0.5,1,0,−1}.

Related Functions   acos, asin, atan

cos, tan

sinh

Summary   This function returns the hyperbolic sine of the specified argument.

Syntax   sinh*(numbers)*

The *numbers* argument can be a scalar or range.

Like the circular trig functions, this function also accepts numbers in degrees, radians, or grads, depending on the units selected in the User-Defined Transform dialog box.

Example   The operation **x** = **sinh**(**col**(3)) sets the variable *x* to be the hyperbolic sine of all data in column 3.

Related Functions   cosh, tanh

sinp

Summary   The sinp function automatically generates the initial parameter estimates for a sinusoidal functions using the FFT method. The three parameter estimates are returned as a vector.

Syntax   sinp(*x range, y range*)

The *x range* argument specifies the x variable, and the y range argument specifies the y variable. Any missing value or text string contained within one of the ranges is ignored and will not be treated as a data point. *x range* and *y range* must be the same size, and the number of valid data points must be greater than or equal to 3.

Σ sinp is especially used to perform smoothing on waveform functions, used in determination of initial parameter estimates for nonlinear regression.

size

Summary   The size function returns a value equal to the total number of elements in the specified range, including all numbers, missing values, and text strings.

Note that $size(X) \frac{1}{2} count(X) + missing(X)$.

Syntax   size*(range)*

The range argument must be a single range (indicated with the { } brackets) or a worksheet column.

Example For Figure 5–1:

the operation **size**(**col**(1)) returns a value of 6,
the operation **size**(**col**(2)) returns a value of 6, and
the operation **size**(**col**(3)) returns a value of 4.

Related Functions count, missing

Figure 5–5

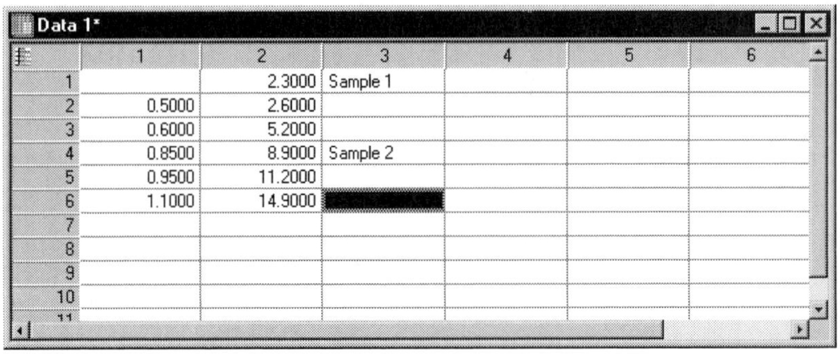

	1	2	3	4	5	6
1		2.3000	Sample 1			
2	0.5000	2.6000				
3	0.6000	5.2000				
4	0.8500	8.9000	Sample 2			
5	0.9500	11.2000				
6	1.1000	14.9000				
7						
8						
9						
10						

sort

Summary This function can be used to sort a range of numbers in ascending order, or a range of numbers in ascending order together with a block of data.

Syntax sort*(**block**,**range**)*

The ***range*** argument can be either a specified range (indicated with the {} brackets) or a worksheet column. If the block argument is omitted, the data in range is sorted in ascending order.

Example The operation **col**(2) = **sort**(**col**(1)) returns the contents of column 1 arranged in ascending order and places it in column 2.

To reverse the order of the sort, you can create a custom function:

reverse(x) = x[data(size(x),1)]

then apply it to the results of the sort. For example, **reverse**(**sort**(**x**)) sorts range *x* in descending order.

<dl>
<dt>Example</dt>
<dd>

The operation:

block(3,1) = sort(block(1,1,2,size(col(2)),col(2))

sorts data in columns 1 and 2 using column 2 as the key column and places the sorted data in columns 3 and 4.
</dd>
</dl>

Related Functions size, data

sqrt

<dl>
<dt>Summary</dt>
<dd>

The sqrt function returns a value or range of values consisting of the square root of each value in the specified range. Numerically, this is the same as {*numbers*}^0.5, but uses a faster algorithm.
</dd>

<dt>Syntax</dt>
<dd>

sqrt*(numbers)*

The *numbers* argument can be a scalar or range of numbers. Any missing value or text string contained within a range is ignored and returned as the string or missing value.

For numbers < 0, sqrt generates a missing value.
</dd>

<dt>Example</dt>
<dd>

The operation **sqrt**({−**1,0,1,2**}) returns the range {--,0,1,1.414}.
</dd>
</dl>

stddev

<dl>
<dt>Summary</dt>
<dd>

The stddev function returns the standard deviation of the specified range, as defined by:

$$s = \left[\frac{1}{n-1} \sum_{i=1}^{n} (x_i - x)^2 \right]^{\frac{1}{2}}$$
</dd>

<dt>Syntax</dt>
<dd>

stddev*(range)*

The *range* argument must be a single range (indicated with the {} brackets) or a worksheet column. Any missing value or text string contained within a range is ignored.
</dd>

<dt>Example</dt>
<dd>

For the range x = {1,2}, the operation **stddev(x)** returns a value of .70711.
</dd>
</dl>

Related Functions stderr

stderr

Summary   The stderr function returns the standard error of the mean of the specified range, as defined by

$$\frac{s}{\sqrt{n}}$$

where s is the standard deviation.

Syntax   stderr*(range)*

The *range* argument must be a single range (indicated with the {} brackets) or a worksheet column. Any missing value or text string contained within a range is ignored.

Example   For the range x = {1,2}, the operation **stderr(x)** returns a value of 0.5.

Related Functions   stddev

subblock

Summary   The subblock function returns a block of cells from within another previously defined block of cells from the worksheet. The subblock is defined using the upper left and lower right cells of the subblock, relative to the range defined by the source block.

Syntax   subblock (*block, column 1, row 1, column 2, row 2*)

The *block* argument can be a variable defined as a block, or a block function statement.

The *column 1* and *row 1* arguments are the relative coordinates for the upper left cell of the subblock with respect to the source block. The *column 2* and *row 2* arguments are the relative coordinates for the lower right cell of the subblock. All values within this range are returned. Operations performed on a block always return a block. If column 2 and row 2 are omitted, then the last row and/or column is assumed to be the last row and column of the source block.

All column and row arguments must be scalar (not ranges).

Example   For x = **block (3,1,20,42)** the operation **subblock (x,1,1,1,1)** returns cell (3,1) and the operation **subblock (x,5,5)** returns the block from cell (7, 5) to cell (20, 42).

Related Functions   block, blockheight, blockwidth

sum

Summary   The function sum returns a range of numbers representing the accumulated sums along the list. The value of the number is added to the value of the preceding cumulative sum.

Because there is no preceding number for the first number in a range, the value of the first number in the result is always the same as the first number in the argument range.

Syntax   sum*(range)*

The **range** argument must be a single range (indicated with the {} brackets) or a worksheet column. Any text string or missing value contained within the range is returned as the string or missing value.

Example   For x = {2,6,7}, the operation **sum(x)** returns a value of {2,8,15}.
For y = {4,12,−6}, the operation **sum(y)** returns a value of {4,16,10}.

Related Functions   diff, total

tan

Summary   This function returns ranges consisting of the tangent of each value in the argument given.

This and other trigonometric functions can take values in radians, degrees, or grads. This is determined by the Trigonometric Units selected in the User-Defined Transform dialog box.

Syntax   tan*(numbers)*

The *numbers* argument can be a scalar or range.

If you regularly use values outside of the usual -2π to 2π (or equivalent) range, use the mod function to prevent loss of precision. Any missing value or text string contained within a range is ignored and returned as the string or missing value.

Example   If you choose Degrees as your Trigonometric Units in the transform dialog box, the operation **tan({0,45,135,180})** returns values of {0,1,−1,0}.

Related Functions   acos, asin, atan
cos, sin

tanh

Summary
:   This function returns the hyperbolic tangent of the specified argument.

Syntax
:   tanh(*numbers*)

    The *numbers* argument can be a scalar or range.

    Like the circular trig functions, this function also accepts numbers in degrees, radians, or grads, depending on the units selected in the User-Defined Transform dialog box.

Example
:   The operation **x** = **tanh(col(3))** sets the variable *x* to be the hyperbolic tangent of all data in column 3.

Related Functions
:   cosh, sinh

total

Summary
:   The function total returns a single value equal to the total sum of all numbers in a specified range. Numerically, this is the same as the last number returned by the sum function.

Syntax
:   total(*range*)

    The *range* argument must be a single range (indicated with the {} brackets) or a worksheet column. Missing values and text strings contained within the range are ignored.

Examples
:   For x = {9,16,7}, the operation **total(x)** returns a value of 32.
    For y = {4,12,−6}, the operation **total(y)** returns a value of 10.

Related Functions
:   diff, sum

x25

Summary
:   The x25 function returns value of the x at $y_{min} + \frac{r_{range}}{4}$ in the ranges of coordinates provided, with optional Lowess smoothing. This is typically used to return the x value for the y value at 25% of the distance from the minimum to the maximum of smoothed data for sigmoidal shaped functions.

Syntax   x25(*x range, y range, f*)

The *x range* argument specifies the x variable, and the y range argument specifies the y variable. Any missing value or text string contained within one of the ranges is ignored and will not be treated as a data point. *x range* and *y range* must have the same size, and the number of valid data points must be greater than or equal to 3.

The optional *f* argument defines the amount of Lowess smoothing, and corresponds to the fraction of data points used for each regression. *f* must be greater than or equal to 0 and less than or equal to 1. $0 \leq f \leq 1$. If *f* is omitted, no smoothing is used.

Example   For x = {0,1,2}, y={0,1,4}, the operation

col(1)=x25(x,y)

places the x at $y_{min} + \frac{r_{range}}{4}$ as 1.00 into column 1.

Related Functions   x50, x75, xatymax, xwtr

x50

Summary   The x50 function returns value of the x at $y_{min} + \frac{r_{range}}{2}$ in the ranges of coordinates provided, with optional Lowess smoothing. This is typically used to return the x value for the y value at 50% of the distance from the minimum to the maximum of smoothed data for sigmoidal shaped functions.

Syntax   x50(*x range, y range, f*)

The *x range* argument specifies the x variable, and the y range argument specifies the y variable. Any missing value or text string contained within one of the ranges is ignored and will not be treated as a data point. *x range* and *y range* must have the same size, and the number of valid data points must be greater than or equal to 3.

The optional *f* argument defines the amount of Lowess smoothing, and corresponds to the fraction of data points used for each regression. *f* must be greater than or equal to 0 and less than or equal to 1. $0 \leq f \leq 1$. If *f* is omitted, no smoothing is used.

Example   For x = {0,1,2}, y={0,1,4}, the operation

col(1)=x50(x,y)

places the x at $y_{min} + \frac{r_{range}}{2}$ as 1.00 into column 1.

Related Functions   x25, x75, xatymax, xwtr

x75

Summary   The x75 function returns value of the x at $y_{min} + \frac{3r_{range}}{4}$ in the ranges of coordinates provided, with optional Lowess smoothing. This is typically used to return the x value for the y value at 75% of the distance from the minimum to the maximum of smoothed data for sigmoidal shaped functions.

Syntax   x75(*x range, y range, f*)

The *x range* argument specifies the x variable, and the y range argument specifies the y variable. Any missing value or text string contained within one of the ranges is ignored and will not be treated as a data point. *x range* and *y range* must have the same size, and the number of valid data points must be greater than or equal to 3.

The optional *f* argument defines the amount of Lowess smoothing, and corresponds to the fraction of data points used for each regression. *f* must be greater than or equal to 0 and less than or equal to 1. $0 \leq f \leq 1$. If *f* is omitted, no smoothing is used.

Example   For x = {0,1,2}, y={0,1,4}, the operation

col(1)=x75(x,y)

places the x at $y_{min} + \frac{3r_{range}}{4}$ as 2.00 into column 1.

Related Functions   x25, x50, xatymax, xwtr

xatymax

Summary   The xatymax function returns the x value at the maximum y value found, with optional Lowess smoothing.

Syntax   xatymax(*x range, y range, f*)

The *x range* argument specifies the x variable, and the y range argument specifies the y variable. Any missing value or text string contained within one of the ranges is ignored and will not be treated as a data point. *x range* and *y range* must have the same size, and the number of valid data points must be greater than or equal to 3.

The optional *f* argument defines the amount of Lowess smoothing, and corresponds to the fraction of data points used for each regression. *f* must be greater than or equal to 0 and less than or equal to 1. $0 \leq f \leq 1$. If *f* is not defined, no smoothing is used.

Σ   If duplicate y maximums are found xatymax will return the average value of all the x at y maximums.

Example   For x = {0,1,2}, y={0,1,4}, the operation

col(1)=xatymax(x,y)

places the x at the y maximum as 2.00 into column 1.

Related Functions   x25, x50, x75, xwtr

xwtr

Summary   The xwtr function returns value of x75-x25 in the ranges of coordinates provided, with optional Lowess smoothing.

Syntax   xwtr(*x range, y range, f*)

The *x range* argument specifies the x variable, and the y range argument specifies the y variable. Any missing value or text string contained within one of the ranges is ignored and will not be treated as a data point. *x range* and *y range* must have the same size, and the number of valid data points must be greater than or equal to 3.

The optional *f* argument defines the amount of Lowess smoothing, and corresponds to the fraction of data points used for each regression. *f* must be greater than or equal to 0 and less than or equal to 1. $0 \leq f \leq 1$. If *f* is omitted, no smoothing is used.

Example   For x = {0,1,2}, y={0,1,4}, the operation

col(1)=xwtr(x,y)

places the x75-x25 as double 1.00 into column 1.

Related Functions   x25, x50, x75, xatymax

User-defined Functions

You can create any user-defined function, consisting of any expression in the transform language, and then refer to it by name.

For example, the following transform defines the function *dist2pts*, which returns the distance between two points

dist2pts(x1,y1,x2,y2) = sqrt((x2−x1)^2+(y2−y1)^2)

You can then use this custom-defined function, instead of the expression to the right of the equal sign, in subsequent equations. For example, to plot the distances between two sets of XY coordinates, with the first points stored in columns 1 and 2, and the second in columns 3 and 4, enter:

col(5) = dist2pts(col(1),col(2),col(3),col(4))

The resulting distances are placed in column 5.

Saving User-Defined Functions

Frequently used variable values and custom transforms can be saved to a transform file, then copied and pasted into the desired transform.

To save user-defined functions to a file, then apply them to a transform:

1. Define the variables and functions in the Transform window, then click the Save button.

2. When the Save dialog box appears, name the file something like "User-Defined Functions."

3. Select the function you want to use in the transform, then press Ctrl+C or Ctrl+Ins.

4. Open the transform file you want to copy the function to, click the point in the text where you want to enter the function, then press Ctrl+V or Shift+Ins.

Example Transforms

Many mathematical transform examples, along with appropriate graphs and worksheets are included with SigmaPlot. This chapter is describes the data transform examples and the graphing transform examples provided. Each description contains the text of the transform and, where applicable, a graph displaying the possible results of the transform.

The sample transforms and the XFMS.JNB notebook can be found in the XFMS folder.

Data Transform Examples

The data transform examples are provided to show you how transform equations can manipulate and calculate data.

One Way Analysis of Variance (ANOVA)

A One Way Analysis of Variance (ANOVA) table can be created from the results of a regression or nonlinear regression. The original Y values, the Y data from the fitted curve, and the parameters are used to generate the table.

The transform assumes you have placed the original Y data in column 2, the fitted Y data in column 3, and the regression coefficients or function parameters in column 4. You can either place this data in these columns, or change the column numbers used by the transform.

The One Way ANOVA transform contains examples of the following transform functions:

➤ count
➤ if
➤ total
➤ mean
➤ {...} (constructor notation)

To use the One Way ANOVA transform:

1. Make sure your original Y data is in column 2. Perform the desired regression using the Regression Wizard, and save your Predicted values (fitted Y data) in column 3, and Parameters (the regression coefficients) in column 4.

   For more information on using the Regression Wizard, see Chapter 8, "Regression Wizard."

2. Press F10 to open the User-Defined Transform dialog box, then click the Open button and open the ANOVA.XFM transform file in the XFMS directory. The ANOVA transform appears in the edit window.

3. Click Run. The ANOVA results are placed in columns 5 through 9, or beginning at the column specified with the anova variable.

One Way
ANOVA Transform
(ANOVA.XFM)

```
'****** Analysis of Variance (ANOVA) Table ******
'This transform takes regression or curve fit
'results and constructs an ANOVA table
'Required INPUT: y data, fitted y data, function
'          parameters/coefficients
'RESULTS: sum of squares, degrees of freedom, mean
'      squared, F-value, R-squared & R values,
'      standard error of fit;
'INPUT to be placed in (specify source columns):
y_col=2      'y data column number
fit_col=3     'fitted y data column number
param_col=4    'parameter column number
'ANOVA to be placed in column:
anova=5       'ANOVA table starting column
          '(5 columns x 10 rows)
y=col(y_col)    'define y values
f=col(fit_col)   'define fitted y values
p=col(param_col) 'define function parameters
n=count(y)      'number of y data points
tdof=n-1       'total degrees of freedom
r=count(if(p<>0,p,"--")) 'the number of nonzero
                'parameters
'******* ANOVA TABLE CALCULATION *******
'Regression Degrees of Freedom:
rdof=tdof-if(r<count(p),r-2,count(p)-1)
'Error Degrees of Freedom:
edof=tdof-rdof
'Sum of Squares of Residuals:
SSE=total((y-f)^2)
'Sum of Squares of Error about the Mean:
SSM=total((y-mean(y))^2)
'Sum of Squares of Error due to Regression:
SSR=SSM-SSE
'Standard Error of Fit:
```

```
se=sqrt(SSE/rdof)
'F value:
f1=((SSM-SSE)/(edof))/(SSE/rdof)
F=if(n<2,"n < 2 !",f1)
'R squared:
R2=(1-SSE/SSM)
'****** PLACE ANOVA TABLE IN WORKSHEET *******
col(anova)={0/0,"REGRESSION","ERROR","TOTAL"}
col(anova+1)={"SUM OF SQUARES",SSR,SSE,SSM}
col(anova+2)={"DEG FREEDOM",edof,rdof,tdof}
col(anova+3)={"MEAN SQUARE",(SSR/edof),(SSE/rdof)}
col(anova+4)={"F",F}
col(anova,7)={"#POINTS","R SQUARED","R","STD ERR"}
col(anova+1,7)={n,R2,sqrt(R2),se}
```

Area Beneath a Curve Using Trapezoidal Rule

This transform computes the area beneath a curve from X and Y data columns using the trapezoidal rule for unequally spaced X values. The algorithm applies equally well to equally spaced X values.

This transform uses an example of the diff function.

To use the Area Under Curve transform:

1. Place your X data in column 1 and your Y data in column 2. If your data has been placed in other columns, you can specify these columns after you open the AREA.XFM file. You can use an existing or new worksheet.

2. Press F10 to open the User-Defined Transform dialog box, then click the Open button and open the AREA.XFM transform file in the XFMS directory. The Area transform appears in the edit window.

3. Click Run. The area is placed in column 3 or in the column specified with the res variable.

Area Under Curve Transform (AREA.XFM)

```
'*Transform for Calculating Area Beneath a Curve*
' This transform integrates under curves using the
'  trapezoidal rule. This can be used for equal
'  or unequally spaced x values.
'  The algorithm is: sigma i from 0 to n-1, or
'  {yi(xi+1 - xi) + (1/2)(yi+1 - yi)(xi+1 - xi)}
' Place your x data in x_col and y data in y_col or
' change the column numbers to suit your data.
' Results are placed in column res.
x_col=1        'column number for x data
y_col=2        'column number for y data
res=3          'column number for result
'Define x and y data
```

```
x=col(x_col)
y=col(y_col)
'*************** CALCULATE AREA ***************
'Compute the range of differences between
' x[i] & x[i-1]
xdif1=diff(x)
n=count(x)          'Delete first value-
xdif=xdif1[data(2,n)]   'not a difference
'Compute the range of differences between
' y[i] & y[i-1]
ydif1=diff(y)        'Delete first value-
ydif=ydif1[data(2,n)]   'not a difference
'Use only y values from y[1] to y[n-1]
y1=y[data(1,n-1)]
'Calculate trapezoidal integration
intgrl=y1*xdif+0.5*ydif*xdif
a=total(intgrl)
'******** PLACE RESULTS IN WORKSHEET *********
col(res)=a     'Put area in column res
```

Bivariate Statistics

This transform takes two data columns of equal length and computes their means, standard deviations, covariance, and correlation coefficient. The columns must be of equal length.

The Bivariate transform uses examples of these transform functions

➤ mean

➤ stddev

➤ total

To use the Bivariate transform:

1. Place your X data in column 1 and your Y data in column 2. If your data has been placed in other columns, you can specify these columns after you open the BIVARIAT.XFM transform file. You can enter data into an existing worksheet or a new worksheet.

2. Press F10 to open the User-Defined Transform dialog box, then click the Open button, and open the BIVARIAT.XFM transform file in the XFMS directory. The Bivariate Statistics transform appears in the edit window.

3. Click Run. The results are placed in columns 3 and 4, or beginning in the column specified with the res variable.

Bivariate Statistics Transform (BIVARIAT.XFM)

```
' This transform integrates under curves using the
' trapezoidal rule
'*** Transform to Compute Bivariate Statistics ***
' This transform takes x and y data and returns
' the means, standard deviations, covariance, and
' correlation coefficient (rxy)
' Place your x data in x_col and y data in y_col
' or change the column numbers to suit your data.
' Results are placed in columns res and res+1
x_col=1     'column number for x data
y_col=2     'column number for y data
res=3       'first results column
'Define x and y data
x=col(x_col)
y=col(y_col)
'************* CALCULATE STATISTICS *************
n=size(col(x_col))   'number of x values
            'n must be > 1
mx=mean(x)        'mean of x data
my=mean(y)        'mean of y data
sx=stddev(x)       'standard deviation of x data
sy=stddev(y)       'standard deviation of y data
'covariance of x and y
sxy=if(n>1,(total(x*y)-n*mx*my)/(n-1),0)
'correlation coefficient of x and y
rxy=sxy/(sx*sy)
'********* PLACE STATISTICS IN WORKSHEET *********
col(res)={"N","MEAN X","MEAN Y","STD DEV X",
 "STD DEV Y","COVARIANCE","CORR COEFF"}
col(res+1)= if(n>1,{n,mx,my,sx,sy,sxy,rxy},"n<=1")
```

Differential Equation Solving

This transform can be used to solve user-defined differential equations. You can define up to four first order equations, named $fp1(x_1,y_1,y_2,y_3,y_4)$ through $fp4(x_1,y_1,y_2,y_3,y_4)$. **Set any unused equations = 0.**

To solve a first order differential equation:

1. Begin a new worksheet by choosing the File menu New command, then choosing Worksheet; this transform requires a clean worksheet to work correctly.

2. Open the User-Defined Transforms dialog box by selecting the Transforms menu User Defined command, then clicking the Open button, and opening the DIFFEQN.XFM transform file in the XFMS directory. The Differential Equation Solving transform appears in the edit window.

3. Scroll to the Number of Equations section and enter a value for the **neqn** variable. This is the number of equations you want to solve, up to four.

4. Scroll down to the Differential Equations section, and set the **fp1** through **fp4** functions to the desired functions. **Set any unused equations = 0.** If only one first order differential equation is used, then only the fp1 transform equation is used and fp2, fp3, and fp4 are set to 0. For example, if you only wanted to solve the differential equation:

$$\frac{dy_1}{dt} = -ay_1$$

you would enter:

fp1(x,y1,y2,y3,y4) = −a*y1
fp2(x,y1,y2,y3,y4) = 0
fp3(x,y1,y2,y3,y4) = 0
fp4(x,y1,y2,y3,y4) = 0

5. Scroll down to the Initial Values heading and set the **nstep** variable to the number of integration (X variable) steps you want to use. The more steps you set, the longer the transform takes.

6. Set the initial X value **x0**, final X value **x1**, and the Y1 through Y4 values (placed in cells (2,1) through (5,1)). **If you are not using a y_1 value, set that value to zero (0).** For example, for the single equation example above, you could enter:

x0 = 0   ;initial x
x1 = 1   ;final x
cell(2,1) = 1   ;y1 initial value
cell(3,1) = 0   ;y2 initial value

cell(4,1) = 0   ;y3 initial value
cell(5,1) = 0   ;y4 initial value

7. Click Run. The results output is placed in columns 1 through **neqn+1**.

8. To graph your results, create a Line Plot graphing column 1 as your X data and columns 2 through 5 as your Y data.

<table>
<tr><td>Figure 6–1
Differential Equation Graph</td><td>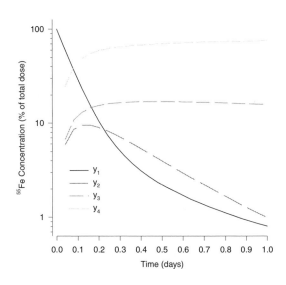</td></tr>
</table>

Plasma Iron Kinetics - IV Bolus ^{55}Fe

For information on creating a graph plotting one X data column against many Y data columns see the SigmaPlot's *User's Manual*.

Differential Equation Solving Transform (DIFFEQN.XFM)

The transform example solves the equations:

$$\frac{dy_1}{dt} = -(r_{65} + r_{75} + r_{85})y_1 + r_{56}y_2 + r_{57}y_3$$

$$\frac{dy_2}{dt} = r_{65}y_1 - r_{56}y_2$$

$$\frac{dy_3}{dt} = r_{75}y_1 - r_{57}y_3$$

$$\frac{dy_4}{dt} = r_{85}y_1$$

```
'*** Solution of Coupled First Order Differential ***
'*** Equations by Fourth Order Runge-Kutta Method ***


'    ***** Number of Equations *****
'Enter the number of differential equations "neqn"
'the number of differential equations, less than or
'equal to 4

neqn = 4      'number of differential equations

'     ***** Differential Equations ******
'Set the functions fp1, fp2, fp3, fp4 to be equal to
'coupled first order ordinary differential equations
'where x is the independent variable and y1, y2, y3,
'and y4 are the dependent variables. The number of
'equations must be equal to the neqn value set above.

'If neqn < 4 then use zeros (0) for the unused
'equations

fp1(x,y1,y2,y3,y4) = -(r65+r75+r85)*y1+r56*y2+r57*y3
fp2(x,y1,y2,y3,y4) = r65*y1-r56*y2
fp3(x,y1,y2,y3,y4) = r75*y1-r57*y3
fp4(x,y1,y2,y3,y4) = r85*y1

'      ****** Initial Values ******
'Enter the maximum number of integration
'steps "nstep".

nstep = 25    ;number of integration steps

'Enter the initial and final x values followed by the
'initial values of y1 (and y2, y3 and y4, if they are
'used). If neqn < 4 then use zeros (0) for the unused
'initial yi values.

x0 = 0        'initial x
x1 = 1        'final x
cell(2,1) = 100 'y1 initial value
cell(3,1) = 0   'y2 initial value
cell(4,1) = 0   'y3 initial value
cell(5,1) = 0   'y4 initial value


'     **** RESULTS ****
' The output will be placed in columns 1 through neqn+1.
'x is placed in column 1. The yi values are placed in
'columns 2 through neqn+1. Other columns are used for
```

```
'program working space.

'    ***** Parameter Values *****
'Enter all necessary parameter values below

r65 = 2.2
r75 = 2.3
r85 = 8.4
r56 = 4.2
r57 = 0.32

'       ********** PROGRAM **********

fp(x,y1,y2,y3,y4,m) = if(m=1, fp1(x,y1,y2,y3,y4),
     if(m=2, fp2(x,y1,y2,y3,y4),
     if(m=3, fp3(x,y1,y2,y3,y4),
     if(m=4, fp4(x,y1,y2,y3,y4)))))
h = (x1-x0)/nstep
hh = 0.5*h
h6 = h/6
cell(1,1) = x0
n2 = neqn+2      'yt
n3 = neqn+3      'dydx
n4 = neqn+4      'dyt
n5 = neqn+5      'dym

' Fixed Step Size Fourth Order Runge-Kutta
for k = 1 to nstep do
  xk = x0 + (k-1)*h
  xh = xk + hh

 for i = 1 to neqn do
   cell(n3,i) = fp(xk,cell(2,k),cell(3,k),
    cell(4,k),cell(5,k),i)   'dydx
   cell(n2,i) = cell(i+1,k) +
    hh*cell(n3,i)            'yt

 end for
 for i1 = 1 to neqn do
  cell(n4,i1) = fp(xh,cell(n2,1),cell(n2,2),
   cell(n2,3),cell(n2,4),i1)   'dyt
  cell(n2,i1) = cell(i1+1,k) +
   hh*cell(n4,i1)           'yt
 end for

 for i2 = 1 to neqn do
  cell(n5,i2) = fp(xh,cell(n2,1),cell(n2,2),
   cell(n2,3),cell(n2,4),i2)   'dym
  cell(n2,i2) = cell(i2+1,k) +
```

```
   h*cell(n5,i2)                'yt
 cell(n5,i2) = cell(n5,i2) +
   cell(n4,i2)          'dym = dym + dyt
end for

for i3 = 1 to neqn do
 cell(n4,i3) = fp(xk+h,cell(n2,1),cell(n2,2),
   cell(n2,3),cell(n2,4),i3)    'dyt
 cell(i3+1,k+1) = cell(i3+1,k) + h6*(cell(n3,i3)
   + cell(n4,i3) + 2*cell(n5,i3))
end for

 cell(1,k+1) = cell(1,k) + h
end for
```

F-test to Determine Statistical Improvement in Regressions

This transform compares two equations from the same family to determine if the higher order provides a statistical improvement in fit.

Often it is unclear whether a higher order model fits the data better than a lower order. Equations where higher orders may produce better fits include: simple polynomials of different order, the sums of exponentials for transient response data, and the sums of hyperbolic functions for saturation ligand binding data.

F-TEST.XFM uses the residuals from two regressions to compute the sums of squares of the residuals, then creates the F statistic and computes an approximate P value for the significance level.

You can try this transform out on the provided sample graph, or run it on the residuals produced by your own regression sessions. Residuals are saved to the worksheet by the Regression Wizard.

1. **To use the provided sample data and graph**, open the F-test worksheet and graph in the XFMS.JNB notebook. The worksheet contains raw data in columns 1 and 2, and curve fit results for the two competitive binding models in columns 3-5 and 6-8. The graph plots the raw data and the two curve fits.

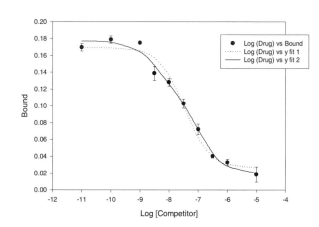

Figure 6–2
Comparing Two Curve
Fits

Use of the F Test

2. **To use your own data**, enter the XY data to be curve fit in columns 1 and 2, respectively. Select the first curve fit equation and use it to fit the data, place the parameters, fit results and residuals in the first empty columns (3-5). Run the second curve fit and place the results in columns 6-8 (the default). If desired, create graphs of these results using the wizard.

3. Press F10 to open the User-Defined Transform dialog box, then open the F-TEST.XFM transform file. Specify n1 and n2, the number of parameters in the lower and higher order functions. In the example provided, these are 3 and 5, respectively.

   If necessary, specify cs1 and cs2, the column locations for the residuals of each curve fit, and cres, the first column for the two column output.

4. Click Run. The *F*-test value and corresponding *P* value are placed into the worksheet. If $P < 0.05$, you can predict that the higher order equation provides a statistically better fit.

F-test Transform (F-TEST.XFM)

```
'***** Compare Two Nonlinear Curve Fits *****
'*****     with the F test       *****
'This transform uses the residuals from two
'curve fits of functions from the same family
'to determine if there is a significant
'improvement in the fit provided by the higher
'order fitting function.
'The F statistic is computed and used to obtain
'an approximate P value.
'****** Input ******
n1=3   'number parameters for 1st function
      '(fewest parameters)
```

```
n2=5   ' number parameters for 2nd function
cs1 = 6   ' residual column for function 1
cs2 = 9   ' residual column for function 2
cres =10   ' first column of two results columns
'****** Program ******
N=size(col(cs1))
ss1=total(col(cs1)^2)
ss2=total(col(cs2)^2)
F = ((ss1-ss2)/ss2)*((N-n2)/(n2-n1))
'Approximate P value for F distribution
N1=n2-n1      ' A&S, Eq. 26.6.15, p. 947
N2=N-n2
x=(F^(1/3)*(1-2/(9*N2))-(1-2/(9*N1)))/
 sqrt(2/(9*N1)+F^(2/3)*2/(9*N2))
'Normal distribution approximation for P value
pi=3.1415926   ' A&S, Eq. 26.2.17, p 932
z=exp(-x^2/2)/sqrt(2*pi)
t=1/(1+.2316419*x)
p=z*(.31938153*t-.356563782*t^2+1.781477937
 *t^3-1.821255978*t^4+1.330274429*t^5)
'****** Output ******
col(cres)={" F = ", " p = "}
col(cres+1)={F,p}
```

R^2 for Nonlinear Regressions

You can use this transform to compute the coefficient of determination (R^2) for the results of a nonlinear regression. The original Y values and the Y data from the fitted curve are used to calculate R^2.

To save the fitted Y values of the nonlinear regression to the worksheet, use the Regression Wizard to save the Function results to the appropriate column (for this transform, column 3).

1. Place your original Y data in column 2 of the worksheet and the fitted Y data in column 3. If your data has been placed in other columns, you can specify these columns after you open the R2.XFM transform file. You can enter data into an existing or a new worksheet.

2. Press F10 to open the User-Defined Transform dialog box, then click the Open button and open the R2.XFM transform file in the XFMS directory. The R^2 transform appears in the edit window.

3. Click Run. The R^2 value is placed in column 4 of the worksheet, or in the column specified with the res variable.

**R Squared
Transform (R2.XFM)**

```
'***Transform to Compute R Square (Coefficient ***
'** of Determination) for Nonlinear Curve Fits **
' Place your y data in y_col and the fitted y data
' in fit_col or change the column numbers to suit
' your data. Results are placed in column res.
y_col=2         'column number for y data
fit_col=3       'column number for fit results
res=4           'column number for R2 result
'Define y and fitted y values
y=col(y_col)
yfit=col(fit_col)
'************** CALCULATE R SQUARE **************
n=yfit-y
d=y-mean(y)
r2=1.0-total(n^2)/total(d^2)
'****** PLACE R SQUARE VALUE IN WORKSHEET *******
col(res)={"R SQUARE",r2}
```

Standard Deviation of Linear Regression Parameters

This transform computes linear 1st-order regression parameter values (slope and intercept) and their standard deviations using X and Y data sets of equal length.

To calculate 1st-order regression parameters and their standard deviations for XY data points:

1. Place the X data in column 1 of the worksheet and the Y data in column 2. If your data is in other columns, you can specify these columns after you open the STDV_REG.XFM transform file. You can enter data into an existing worksheet or a new worksheet.

2. Press F10 to open the User-Defined Transform dialog box, then click the Open button, and open the STDV_REG.XFM transform file in the XFMS directory. If necessary, change the x_col, y_col, and res variables to the correct column numbers.

3. Click Run. The results are placed in columns 3 and 4, or in the columns specified by the res variable.

**Standard Deviation
Regression
Transform
(STDV_REG.XFM)**

```
'** Transform to Compute Standard Deviations of **
'******** Linear Regression Coefficients *********'
'Place your x data in x_col and y data in y_col or
' change the column numbers to suit your data.
' Results are placed in columns res and res+1.
x_col=1         'column number for x data
y_col=2         'column number for y data
res=3       'first results column
x=col(x_col)    'Define x values
y=col(y_col)    'Define y values
```

```
'********** CALCULATE PARAMETER VALUES **********
n=size(col(x_col))   'n must be > 2
mx=mean(x)
my=mean(y)
sumxx=total(x*x)
sumyy=total(y*y)
sumxy=total(x*y)
a1=(sumxy-n*mx*my)/(sumxx-n*mx*mx)   'slope
a0=my-a1*mx                'intercept
'**** CALCULATE PARAMETER STANDARD DEVIATIONS ****
sregxy=if(n>2,sqrt((sumyy-n*a0*my-a1*sumxy)/(n-2)) ,0)
s0=sregxy*sqrt(sumxx/(n*(sumxx-n*mx*mx))) 'SD a0
s1=sregxy/sqrt(sumxx-n*mx*mx)        'SD a1
'********** PLACE RESULTS IN WORKSHEET **********
col(res)={"n","INTERCEPT","SLOPE","STD DEV INT ",
 "STD DEV SLOPE"}
col(res+1)=if(n>2,{n,a0,a1,s0,s1},{"n <= 2"})
```

Graphing Transform Examples

The graph transform examples are provided to show you how transform equations can manipulate and calculate data to create complex graphs.

Each of the following descriptions provide instructions on how to use SigmaPlot to create graphs. Most of these graphs, however, are already set up as sample graphs. If you use the provided worksheet and graphs with the corresponding transform files, SigmaPlot will automatically create the graphs after you run the transform.

Control Chart for Fractional Defectives with Unequal Sample Sizes

This example computes the fraction of defectives *p* for a set of unequally sized samples using their corresponding numbers of defects, the control limits for *p*, and data for the upper and lower control lines.

This transform contains examples of the following transform functions:

➤ stddev

➤ sqrt

To calculate and graph the fraction of defectives and control lines for given sample sizes and number of defects per sample, you can either use the provided sample data and graph or begin a new notebook, enter your own data and create your own graph using the data.

1. To use the provided sample data and graph, open the Control Chart worksheet and graph in the Control Chart section of the Transform Examples notebook. The worksheet appears with data in columns 1, 2, and 3. The graph page appears with an empty graph.

2. To use your own data, place the sample sizes in column 1 and the corresponding number of defects data in column 2 of a new worksheet. If your data is in other columns, you can specify these columns after you open the CONTCHRT.XFM transform file. You can enter your data in an existing or a new worksheet.

3. Press F10 to open the User-Defined Transform dialog box, then click the Open button and open the CONTCHRT.XFM transform file in the XFMS directory. The Control Chart transform appears in the edit window.

4. Click Run. The results are placed in columns 4 through 5 of the worksheet.

5. If you opened the Control Chart graph, view the graph page. The graph plots the fraction of defectives using a Line and Scatter plot with a Simple Straight Line style graphing column 3 as Y data versus the row numbers. The control lines are plotted as a Simple Horizontal Step Plot using columns 4 and 5 versus their row numbers. The mean line for the fractional defectives is drawn with a reference line.

6. To create your own graph, create a Line and Scatter Plot, with a Simple Line style, then plot column 3 as Y data against the row numbers. Add an additional Line Plot using the Multiple Horizontal Step Plot style, plotting columns 4 and 5 versus their two numbers, then add a reference line to plot the mean line for the fractional device.

   For more information on creating graphs in SigmaPlot, see "Creating and Modifying Graphs" on page 143 in the SigmaPlot *User's Manual.*

Figure 6–3
Control Chart Graph

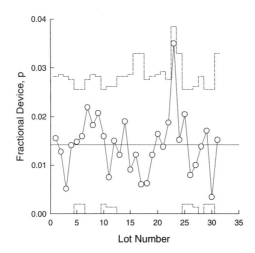

Control Chart
Transform
(CONTROL.XFM)

```
;Transform for Fraction Defective Control Chart
;with Unequal Sample Sizes
; This transform takes sample size and number of
; defectives returns the fraction of defectives
; and data for upper and lower control limit lines
; Place the sample size data in n_col and the
; number of defects in def_col, or change the
; column numbers to suit your data. Fraction
; defective results are placed in the percent_col
; column, and the data for the control lines is
; placed in columns cl and cl+1.
n_col=1     'sample size column
def_col=2   'number of defectives column
percent_col=3 'fraction defective results column
cl=4        'first results column for control
       'limit line data
'********* CALCULATE FRACTION DEFECTIVE **********
n=col(n_col)         'sample sizes
def=col(def_col)        'defectives
col(p_col)=def/n        'fraction defective
pbar=total(def)/total(n)
'*********** CALCULATE CONTROL LIMITS ************
stddev = sqrt(pbar*(1 - pbar)/n)
ucl=pbar+3*stddev   'upper control limit
lcl=pbar-3*stddev   'lower control limit
lclt=if(lcl>=0,lcl,0) 'truncated lower control
             'limit
'**** DATA FOR CONTROL LIMIT LINE STEP CHART ****
col(cl)=ucl
col(cl+1)=lclt
```

Cubic Spline Interpolation and Computation of First and Second Derivatives

This example takes data with irregularly spaced X values and generates a cubic spline interpolant. The CBESPLN1.XFM transform takes X data which may be irregularly spaced and generates the coefficients for a cubic spline interpolant. The CBESPLN2.XFM transform takes the coefficients and generates the spline interpolant and its two derivatives.

The values for the interpolant start at a specified minimum X which may be less than, equal to, or greater than the X value of the original first data point. The interpolant has equally spaced X values that end at a specified maximum which may be less than, equal to, or greater than the largest X value of the original data.

Note that this is not the same algorithm that SigmaPlot uses; this algorithm does not handle multiple valued functions, whereas SigmaPlot does.

To use the transform to generate and graph a cubic spline interpolant, you can either use the provided sample data and graph, or begin a new notebook, enter your own data and create your own graph using the data.

1. To use the provided sample data and graph, open the Cubic Spline worksheet and graph by double-clicking the graph page icon in the Cubic Spline section of the Transform Examples notebook. The worksheet appears with data in columns 1 and 2 and the graph page appears with two graphs. The first graph plots the original XY data as a scatter plot. The second graph appears empty.

2. To use your own data, enter the irregularly spaced XY data into the worksheet. The X values must be sorted in strictly increasing values. The default X and Y data columns used by the transform are columns 1 and 2, respectively.

3. Press F10 to open the User-Defined Transform dialog box, then click the Open button, and open the CBESPLN1.XFM transform file in the XFMS directory. The first Cubic Spline transform appears in the edit window.

4. Move to the Input Variables heading. Set the X data column variable **cx**, the Y data column **cy**, the beginning interpolated X value **xbegin**, the ending interpolated X value **xend**, and the X increments for the interpolated points **xstep**. A larger X step results in a smoother curve but takes longer to compute.

5. Enter the end condition setting **iend** for the interpolation.

6. You can use first, second, or third order conditions.

   If you have only a few data points, you should try different orders to see which one you like the most. See the example for the effect of too low an order on the first and second derivatives.

    1   end spline segments approach straight lines asymptotically

    2   end spline segments approach parabolas asymptotically

    3   end spline segments approach cubics asymptotically

7.   Move to the RESULTS heading and enter the first column number for the results **cr**. This column for the beginning of the results block is specified in both transforms.

8.   Click Run to run the transform. When it finishes, press F10 then open the CBESPLN2.XFM transform file in the XFMS directory. Make sure that the **cr** variable is identical to the previous value, then click Run.

9.   If you opened the Cubic Spline graph, view the page. The first graph plots the original XY data as a scatter plot and the interpolated data as a second line plot by picking the cr column as the X column and cr+1 as the Y column. The second graph plots the derivatives as line plots using the cr column versus the cr+2 column and the cr column versus the cr+3 column.

10. To create your own graphs using SigmaPlot, create a Scatter Plot using a Simple Scatter style which plots the original data in columns 1 and 2 as XY pairs. Add an additional Line Plot using a Simple Spline Curve, then plot the cr column as the X column against the cr+1 column as the Y column.

For more information on creating graphs in SigmaPlot, see the SigmaPlot *User's Manual*.

Figure 6–4
Cubic Spline Graph

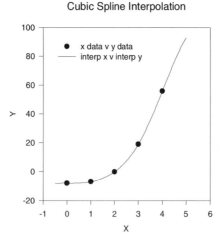

Cubic Spline Interpolation

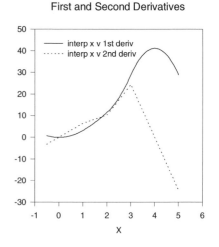

First and Second Derivatives

**Cubic Spline
1 Transform
(CBESPLN1.XFM)**

```
'**** Cubic Spline Interpolation and Computation ****
'       **** of Derivatives ****

'This transform takes an x,y data set with increasing
'ordered x values and computes a cubic spline
'interpolation. The first and second derivatives of
'the spline are also computed.

'Two transform files are run in sequence. This
'transform computes the spline coefficients. The
'CBESPLN2.XFM transform computes the spline and two
'derivatives.
'      ********** Input Variables **********

cx=1        'x data column number
cy=2        'y data column number
xbegin=-.5    'first x value for interpolation
xend=5       'last x value for interpolation
xstep=.025    'x interval for interpolation

'There are 3 spline end conditions allowed:
'   iend = 1: linear end conditions
'   iend = 2: quadratic end conditions
'   iend = 3: cubic end conditions

iend=1       'end condition = 1, 2, or 3

'      *********** RESULTS ***********
'The results are placed into a block of 9 columns
'starting at column cr. Column cr MUST be
'specified identically in both transforms. Columns
'cr to cr+3 contain the x mesh the spline and the
'first two derivatives. Columns cr+4 to cr+7
'contain the a, b, c and d spline coefficients.
'Column cr+8 is for working variables.

cr=3        '1st column of results block

'      *********** PROGRAM ***********
cr4=cr+4    'column for "a" spline coefficients
cr5=cr+5    'column for "b" spline coefficients
cr6=cr+6    'column for "c" spline coefficients
cr7=cr+7    'column for "d" spline coefficients
cr8=cr+8    'working column
n=size(col(cx))
cell(cr8,1)=cx
cell(cr8,2)=cy
cell(cr8,3)=cr
cell(cr8,4)=xbegin
```

```
cell(cr8,5)=xend
cell(cr8,6)=xstep

'compute S for n-2 rows
nm1=n-1
nm2=n-2
cell(cr8,7)=cell(cx,2)-cell(cx,1)        'dx1
cell(cr8,8)=(cell(cy,2)-cell(cy,1))
 /cell(cr8,7)*6                'dy1
for i=1 to nm2 do
 dx2=cell(cx,i+2)-cell(cx,i+1)        'dx2
 dy2=(cell(cy,i+2)-cell(cy,i+1))/dx2*6  'dy2
 cell(cr4,i)=cell(cr8,7)            'dx1
 cell(cr5,i)=2*(cell(cr8,7)+dx2)     '2(dx1+dx2)
 cell(cr6,i)=dx2                 'dx2
 cell(cr7,i)=dy2-cell(cr8,8)          'dy2-dy1
 cell(cr8,7)=dx2                'dx1=dx2
 cell(cr8,8)=dy2                'dy1=dy2
end for

'adjust first and last rows for end condition
dx11=cell(cx,2)-cell(cx,1)
dx1n=cell(cx,n)-cell(cx,nm1)
if iend=2 then
 cell(cr5,1)=cell(cr5,1)+dx11
 cell(cr5,nm2)=cell(cr5,nm2)+dx1n
else if iend = 3 then
 dx12=cell(cx,3)-cell(cx,2)
 cell(cr5,1)=(dx11+dx12)*(dx11+2*dx12)/dx12
 cell(cr6,1)=(dx12*dx12-dx11*dx11)/dx12
 dx2n=cell(cx,nm1)-cell(cx,nm2)
 cell(cr4,nm2)=(dx2n*dx2n-dx1n*dx1n)/dx2n
 cell(cr5,nm2)=(dx1n+dx2n)*(dx1n+2*dx2n)/dx2n
end if
end if

'solve the tridiagonal system
'first reduce
for j = 2 to nm2 do
 jm1=j-1
 cell(cr4,j)=cell(cr4,j)/cell(cr5,jm1)
 cell(cr5,j)=cell(cr5,j)-cell(cr4,j)*cell(cr6,jm1)
 cell(cr7,j)=cell(cr7,j)-cell(cr4,j)*cell(cr7,jm1)
end for

' next back substitute
cell(cr7,nm1)=cell(cr7,nm2)/cell(cr5,nm2)
for k =nm2-1 to 1 step -1 do
  cell(cr7,k+1)=(cell(cr7,k)-cell(cr6,k)*
```

```
        cell(cr7,k+2))/cell(cr5,k)
end for

' specify the end conditions

if iend = 1 then          'linear ends
 cell(cr7,1)=0.0
 cell(cr7,n)=0.0
else if iend = 2 then       'quadratic ends
 cell(cr7,1)=cell(cr7,2)
 cell(cr7,n)=cell(cr7,nm1)
else if iend = 3 then        'cubic ends
 cell(cr7,1)=((dx11+dx12)*cell(cr7,2)-
   dx11*cell(cr7,3))/dx12
 cell(cr7,n)=((dx2n+dx1n)*cell(cr7,nm1)-
   dx1n*cell(cr7,nm2))/dx2n
end if
end if
end if
' compute coefficients of cubic polynomial
for m = 1 to nm1 do
 mp1=m+1
 h=cell(cx,mp1)-cell(cx,m)
 cell(cr4,m)=(cell(cr7,mp1)-cell(cr7,m))/(6*h) 'a(i)
 cell(cr5,m)=cell(cr7,m)/2              'b(i)
 cell(cr6,m)=((cell(cy,mp1)-cell(cy,m))/h)-
   ((2*h*cell(cr7,m)+h*cell(cr7,mp1))/6) 'c(i)
end for
```

Cubic Spline 2 Transform (CBESPLN2.XFM)

```
'********** Spline Generation **********
'Run this transform after you run CBESPLN1.XFM.
'Make sure to enter the same results column
'number value cr as in CBESPLN1.XFM.

'     ********** Input Variables **********

cr=3 '1st column of results block, contains
    'spline x mesh. This must be the same
    'value as in CBESPLN1.XFM.

'     *********** PROGRAM ***********
cr1=cr+1  'column for spline values
cr2=cr+2  'column for 1st derivative of spline
cr3=cr+3  'column for 2nd derivative of spline
cr4=cr+4  'column for "a" spline coefficients
cr5=cr+5  'column for "b" spline coefficients
cr6=cr+6  'column for "c" spline coefficients
cr8=cr+8  'working column
xbegin=cell(cr8,4)
```

```
xend=cell(cr8,5)
xstep=cell(cr8,6)
cx=cell(cr8,1)
cy=cell(cr8,2)

n=size(col(cx))
x1end=int((xend-xbegin)/xstep)+1
cell(cr8,9)=1            'index of x value
x=col(cx)
f(a,b,c,y,dxx)=y+dxx*(c+dxx*(b+dxx*a))
f1(a,b,c,dxx)=c+dxx*(2*b+dxx*(3*a))
f2(a,b,dxx)=2*b+6*a*dxx

for u1 = 1 to x1end do
 u=xbegin+(u1-1)*xstep
 cell(cr,u1)=u         'put u value in col cr
 xj=cell(cr8,9)
 if u <= x[n] then
  if u <= x[xj+1] then   'test u <= x(i+1)
  dx=u-x[xj]          'dx
   cell(cr1,u1)=f(cell(cr4,xj),cell(cr5,xj),
    cell(cr6,xj),cell(cy,xj),dx)
   cell(cr2,u1)=f1(cell(cr4,xj),
    cell(cr5,xj),cell(cr6,xj),dx)
   cell(cr3,u1)=f2(cell(cr4,xj),cell(cr5,xj),dx)
  else
   for j1 = 1 to n do 'start search loop
   if j1>1 then
   if u <= x[j1] then
   if u > x[j1-1] then
   xj1=j1-1
   dx1=u-x[xj1]      'dx
   cell(cr1,u1)=f(cell(cr4,xj1),cell(cr5,xj1),
    cell(cr6,xj1),cell(cy,xj1),dx1)
   cell(cr2,u1)=f1(cell(cr4,xj1),cell(cr5,xj1),
    cell(cr6,xj1),dx1)
   cell(cr3,u1)=f2(cell(cr4,xj1),
    cell(cr5,xj1),dx1)
   cell(cr8,9)=j1-1
   end if
   end if
   end if
   end for        'end search loop
   end if
   else
  xj2=xj
  dx2=u-x[xj2]
  cell(cr1,u1)=f(cell(cr4,xj2),cell(cr5,xj2),
   cell(cr6,xj2),cell(cy,xj2),dx2)
```

```
cell(cr2,u1)=f1(cell(cr4,xj2),cell(cr5,xj2),
  cell(cr6,xj2),dx2)
cell(cr3,u1)=f2(cell(cr4,xj2),cell(cr5,xj2),dx2)
 end if
end for
```

Fast Fourier Transform

The Fast Fourier Transform converts data from the time domain to the frequency domain. It can be used to remove noise from, or smooth data using frequency-based filtering. Use the fft function to find the frequency domain representation of your data, then edit the results to remove any frequency which may adversely affect the original data.

The Fast Fourier Transform uses the following transform functions:

➤ fft
➤ invfft
➤ real
➤ img
➤ complex
➤ mulcpx
➤ invcpx

The Fast Fourier Transform operates on a range of real values or a block of complex values. For complex values there are two columns of data. The first column contains the real values and the second column represents the imaginary values. The worksheet format of a block of complex numbers is:

r_1	i_1
r_2	i_2
....	....
r_n	i_n

where r values are real elements, and i values are imaginary elements. In transform language syntax, the two columns $\{\{r_1, r_2, ... r_n\}, \{i_1, i_2, ... i_n\}\}$ are written as:

```
block({r₁, r₂, ... rₙ},{i₁, i₂, ... iₙ})
```

This function works on data sizes of size 2^n numbers. If your data set is not 2^n in length, the fft function pads 0 at the beginning and end of the data range to make the length 2^n. A procedure for unpadding the results is given in the example *Smoothing with a Low Pass Filter* on page 99.

The fft function returns a range of complex numbers. The Fast Fourier Transform is usually graphed with respect to frequency. To produce a frequency scale, use the relationship:

```
f=fs*(data(0,n/2)-1)/n
```

where *fs* is the sampling frequency. The example transform POWSPEC.XFM. includes the automatic generation of a frequency scale (see page 94).

The *Fast Fourier Transform* operates on data which is assumed to be periodic over the interval being analyzed. If the data is not periodic, then unwanted high frequency components are introduced. To prevent these high frequency components from occurring, *windows* can be applied to the data before using the fft transform. The Hanning window is a cosine function that drops to zero at each end of the data. The example transform POWSPEC.XFM includes the option to implement the Hanning window (see page 94).

Using the Block Function

To return the full fft data to the worksheet:

1. First assign the data you want to filter to column 1 of the worksheet. You can generate the data using a transform, or use your own measurements.

2. Press F10 to open the User-Defined Transforms dialog box, then click the New button to start a new transform.

3. Type the following transform in the edit window:

```
x=col(1)      'real data
tx=fft(x)     'compute the fft
block(2)=tx   'place real fft data back in col(2)
              'place imaginary fft data in col(3)
```

4. Click Run. The results are placed starting one column over from the original data.

Computing Power Spectral Density

The example transform POWSPEC.XFM uses the Fast Fourier Transform function, then computes the power spectral density, a frequency axis, and makes optional use of a Hanning window.

To calculate and graph the power spectral density of a set of data, you can either use the provided sample data and graph, or begin a new notebook, enter your own data and create your own graph using the data.

1. To use the sample worksheet and graph, open the Power Spectral Density worksheet and graph by double-clicking the graph page icon in the Power Spectral Density section of the Transform Examples notebook. Data appears in column 1 of the worksheet, and two graphs appear on the graph page. The top graph shows data generated by the sum of two sine waves plus Gaussian random noise.

The data is represented by:

f(t)=sin(2*pi*f1*t)+0.3*sin(2*pi*f2*t)+g(t)

where f1=10 cycles/sec (cps), f2=100cps, and the Gaussian random noise has mean 0 and standard deviation of 0.2. The lower graph is empty.

2. To use your own data, place your data in column 1. If your data is in a different column, specify the new column after you open the POWSPEC.XFM transform file.

3. Press F10 to open the User-Defined Transform dialog box, then click the Open button, and open POWSPEC.XFM transform file in the XFMS directory. The Power Spectral Density transform appears in the edit window.

$\sum$ To use this transform, the Trigonometric Units must be set to Radians.

4. Click Run. Since the frequency sampling value (fs) is nonzero, a frequency axis is generated in column 2 and the power spectral density data in column 3.

5. If you opened the Power Spectral Density graph, view the graph page. Two graphs appear on the page. The top graph plots the data generated by the sum of two sine waves plus Gaussian random noise using a Line Plot with Simple Straight Line style graphing column 1 versus row numbers. The lower graph plots the power spectral density using a Line Plot with a Simple Straight Line style, graphing column 2 as the X data (frequency), and column 3 as the Y data.

6. To plot your own data using SigmaPlot, choose the Graph menu Create Graph command, or select the Graph Wizard from the toolbar. Create a Line Plot with a Simple Straight Line style plotting your original data versus row numbers by choosing Single Y data format. If you set the frequency sampling value (fs) to nonzero, create a Line Plot with a Simple Straight Line style, graphing columns 2 and 3 using XY Pair data format. Otherwise, create a Line Plot with a Simple Straight Line style plotting column 3 (power spectral density) versus row numbers by choosing Single Y data format.

The power spectral density plot of the signal f(t) shows two major peaks at the two frequencies of the sine waves (10cps and 100cps), and a more or less constant noise level in between.

For more information on how to create graphs in SigmaPlot, see the SigmaPlot *User's Manual.*

Figure 6–5
Power Spectral
Density Example Graph

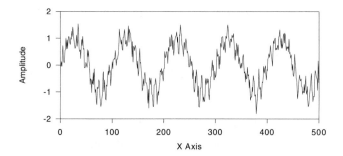

The top graph shows f(t) data generated by the sum of two sine waves plus Gaussian random noise. The bottom graph is the power spectral density of the signal f(t).

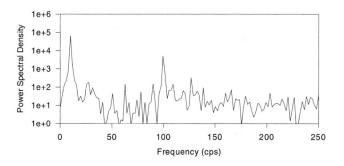

The Power Spectral Density Transform (POWSPEC.XFM)

```
'   This transform computes the power spectral '   density
(psd) of data in column ci and places it
'    in column co.
'    If a nonzero sampling frequency fs is specified
'    then a frequency axis is placed in column co
'    with the psd in the next adjacent column.
'    If han=1 then a Hanning window is applied to the
'    padded data.
'    Set Trigonometric Units to Radians
' Input
ci=1    'input column number
co=2    'first output column number
fs=10   'sampling frequency (produces frequency axis
        ' if fs>0)
han=1   'Hanning window (1=use, 0=don't use)
' Program
pi=3.1415926
x1=col(ci)
if han=1 then       'use Hanning window
 n=size(x1)
 nlog2=log(n)/log(2)   'pad data if necessary
 powup=int(nlog2)
```

```
intup1=if(nlog2-powup<1e-14, 2^powup, 2^(powup+1))
rl=if(mod(n,2)>0, (intup1-n+1)/2, (intup1-n+2)/2
ru=if(mod(n,2)>0, intup1-rl, intup1-rl+1)
x=if(rl-1>0, if(intup1-ru>0, {data(0,0,rl-1), x1,
data(0,0,intup1-ru)},{data(0,0,rl),x1}),
   if(intup1-ru>0, {x1,data,(0,0,intup1-ru)}, {x1}))
w=.5*(1-cos(2*pi*data(0,intup1-1)/(intup1-1)))
 xf=w*x         'multiply padded data by window
else
  xf=x1
end if


tx=fft(xf)           'fft of data
nf=size(tx)/4        'half the zero padded
               'data length
spec=real(tx)^2+img(tx)^2 'power spectral density
spechalf=spec[data(1,nf+1)] 'half the symmetric psd
               'data
f=fs*data(0,nf)/(2*nf)    'frequency axis

' Output
col(co)=if(fs>0,f,spechalf)
col(co+1)=if(fs>0,spechalf)
```

Kernel Smoothing

The example transform SMOOTH.XFM smooths data by convolving the Fast Fourier Transform of a triangular smoothing kernel together with the fft of the data. Smoothing data using this transform is computationally very fast; the number of operations is greatly reduced over traditional methods, and the results are comparable. To increase the smoothing, increase the width of the triangular smoothing kernel.

To calculate and graph the smoothed data, you can either use the provided sample data and graph, or begin a new notebook, enter your own data, and create your own graph using the data.

1. To use the sample worksheet and graph, open the Kernel Smoothing worksheet and graph by double-clicking the graph page icon in the Kernel Smoothing section of the Transform Examples notebook. Data appears in columns 1 through 4, 6, and 7 of the worksheet, and two graphs appear on the graph page. The first graph has two plots, the signal, and the signal with noise distortion. Column 1 contains the X data, column 2 contains the Y data for the signal, and column 3 contains the Y data for the signal and the noise distortion. The lower graph is empty.

2. To use your own data, place your data in columns 1 through 2. If your data is in other columns, specify the new columns after you open the SMOOTH.XFM transform file. If necessary, specify a new column for the results.

3. Press F10 to open the User-Defined Transform dialog box, then click the Open button, and open SMOOTH.XFM transform file in the XFMS directory. The Kernel Smoothing transform appears in the edit window.

Σ To use this transform, make sure the Insert mode is turned off.

4. Click Run. The results are placed in column 5 unless you specified a different column in the transform.

5. If you opened the Kernel Smoothing graph, view the graph page. Two graphs appear on the page. The first graph has two plots, the signal, and the signal with noise distortion. The Line Plot with a Multiple Straight Line style graphs column 1 as the X data, column 2 as the Y data for the signal, and column 3 as the Y data for the signal and the noise distortion. The lower Line Plot with a Simple Straight Line style plots column 1 as the X data, and column 5 as the Y data using XY Pairs data format.

6. To plot your own data using SigmaPlot, choose the Graph menu Create Graph command, or select the Graph Wizard from the toolbar. Create a Line Plot with a Multiple Straight Line style using X Many Y data format, plotting column 1 as the X data, column 2 as the Y data for the signal, and column 3 as the Y data for the signal and the noise distortion. Create a second Line Plot graph with a Simple Straight Line style using the data in columns 1 and 5, graphing column 1 as the X data and column 5 as the Y data using XY Pairs data format.

For more information on how to create graphs in SigmaPlot, see the SigmaPlot *User's Manual.*

Figure 6–6
Kernel Smoothing Graph

The top graph shows two plots: the signal, and the signal plus noise distortion. The bottom graph is the kernel smoothing of the signal with smoothing set at 10%.

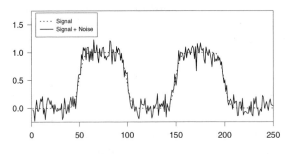

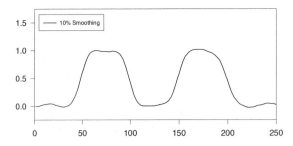

The Kernel Smoothing Transform (SMOOTH.XFM)

```
' Kernel Smoothing
' This transform smooths data using kernel smoothing

' Input
ci=3      ' input column number
co=5      ' output column number
r=10      ' percentage smoothing

' Program
x=col(ci)      ' data
n1=size(x)
tx=fft(x)         ' fft of data
nx=size(tx)/2
n=if(int(r*nx/200)>0, int(r*nx/200), 1)
' generate triangular smoothing kernel
lt={data(n,0,-1), data(0,0,nx-2*n-2), data(0,n)}
lt1=lt/total(lt)
tk=fft(lt1)          ' fft of kernel
td=mulcpx(tk,tx)     ' convolve kernel and data
sd=invfft(td)        ' transform back to time domain
tsd=real(sd)         ' normalize data

' Output
ru=if(mod(n1,2)>0, (nx-n1+1)/2, (nx-n1+2)/2)
' strip out padded channels
rl=if(mod(n1,2)>0, nx-ru, nx-ru+1)
tsd1=tsd[data(ru,rl)]
col(co)=tsd1  ' save smoothed data to worksheet
```

Smoothing with a Low Pass Filter

The Low Pass Filter transform smooths data by eliminating high frequencies. Use this transform in contrast to the Kernel Smoothing transform which smooths data by augmenting some frequencies while minimizing others. The transform statements describing how the low pass filter works are:

```
x=col(1)      'the data to smooth
f=5           'number of channels to eliminate

tx=fft(x)          'fft of data
r=data(1,size(tx)/2) 'total number of channels
mp=size(tx)/4        'get the midpoint
             'remove the frequencies
td=if( r<mp-f or r>mp+1+f,tx,0)
sd=invfft( td )      'convert back to time domain

col(2)=real(sd)      'save smoothed data to worksheet
```

The LOWPASS.XFM transform expresses f as a percentage for ease of use. As the value of f increases, more high frequency channels are removed. Note that this is a digital transform which cuts data at a discrete boundary. In addition, this transform does not alter the phase of the data, which makes it more accurate than analog filtering. A high pass or band pass filter can be constructed in the same manner.

To calculate and graph the smoothing of a set of data using a low pass filter, you can either use the provided sample data and graph, or begin a new notebook, enter your own data, and create your own graph using the data.

1. To use the sample worksheet and graph, open the Low Pass Smoothing worksheet and graph by double-clicking the graph page icon in the Low Pass Smoothing section of the Transform Examples notebook. Data appears in columns 1 through 4 of the worksheet, and two graphs showing plots appear on the graph page. Column 1 contains the X data, column 2 contains the Y data for the signal and the noise distortion, column 3 contains the X data, and column 4 contains the Y data for the original signal. The top graph plots the signal plus the noise distortion; the bottom graph plots the signal.

2. To use your own data, place your data in columns 1 through 2. If your data is in other columns, specify the new columns after you open the LOWPASS.XFM transform file. If necessary, specify a new column for the results.

3. Press F10 to open the User-Defined Transform dialog box, then click the Open button, and open LOWPASS.XFM transform file in the XFMS directory. The Low Pass Filter transform appears in the edit window.

Σ To use this transform, make sure Insert mode is turned off.

4. Click Run. The results are placed starting in column 5, unless you specified a different column in the transform.

5. If you opened the Low Pass Smoothing graph, view the graph page. Two graphs appear. The top graph plots the signal plus the noise distortion, using a Line Plot with a Simple Straight Line style and XY Pairs data format graphing column 1 as the X data, column 2 as the Y data for the signal and the noise distortion. The bottom graph displays two plots. A Scatter Plot with a Simple Scatter Style and XY Pairs data format, plots column 3 as the X data, and column 4 as the Y data for the original signal. A second Line Plot with a Simple Straight Line style using data in columns 1 and 5, plots column 1 as the X data and column 5 as the Y data using XY Pairs data format.

6. To plot your own data using SigmaPlot, choose the Graph menu Create Graph command, or select the Graph Wizard from the toolbar. Create two graphs. Graph the signal plus the noise distortion, using a Line Plot with a Simple Straight Line style and XY Pairs data format graphing column 1 as the X data, column 2 as the Y data for the signal and the noise distortion. Create a second graph with two plots. Plot the original signal using a Scatter Plot with a Simple Scatter Style and XY Pairs data format, plotting column 3 as the X data, and column 4 as the Y data for the original signal. Add a second Line Plot with a Simple

Straight Line style using data in columns 1 and 5, plotting column 1 as the X data and column 5 as the Y data using XY Pairs data format.

For more information on how to create graphs in SigmaPlot, see the SigmaPlot *User's Manual.*

Figure 6–7
Low Pass Filter
Smoothing Graph

The top graph shows the signal plus noise distortion. The bottom graph shows the signal and the low pass filtering set at 88%.

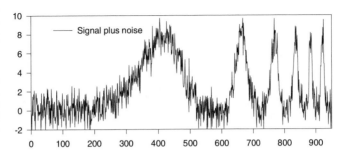

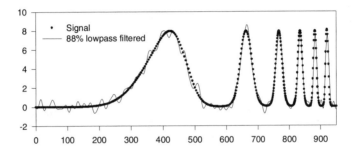

Low Pass Filter
Transform
(LOWPASS.XFM)

```
' Lowpass Smoothing Filter
'   This transform computes the fft, eliminates
'   specified high frequencies and computes the
'   inverse fft
' Input
ci =2              ' input data column number
co = 5             ' output lowpass filtered data
            ' column number
pr = 88            ' % high frequencies to remove
            ' (0-100)
' Program
x = col(ci)
n=size(x)
pr1=if(pr<0,0,if(pr>100,100,pr)) ' trap input pr
                    ' errors
f=int((pr1/100)*mp)    'number of channels to
                ' eliminate
tx = fft(x)        ' fft of data
```

```
r = data(1,size(tx)/2) ' number of data + padded
                    ' channels
mp = size(tx)/4     ' mid point of symmetric
                    ' channels
fc = if( r<mp-f+1 or r>mp+f ,1,0 )  ' eliminate high
                         ' frequencies
td = mulcpx(complex(fc),tx)
sd = invfft( td )    ' convert back to time domain

' Output
ru=if(mod(n,2)>0, (2*mp-n+1)/2, (2*mp-n+2)/2)
               ' remove padded channels
rl=if(mod(n,2)>0, 2*mp-ru, 2*mp-ru+1)
col(co) = real(sd)[data(ru,rl)]
               ' place results in worksheet
block(6,1)=tx
cell(8,1)= mp
cell(8,2)=f
cell(8,3)=pr1
cell(8,4)=n
col(9)=fc
col(10)=r
col(11)=real(tx)^2+img(tx)^2   ' PSD
```

Gain Filter Smoothing

The GAINFILT.XFM transform example demonstrates gain filter smoothing. This method eliminates all frequencies with power spectral density levels below a specified threshold. The transform statements describing how gain filter smoothing works are:

```
P=4000       'psd threshold
x=col(1)      'data

tx=fft(x)         'compute fft of data
md=real(tx)^2+img(tx)^2  'compute sd
kc=if(md>P,1,0)       'remove frequencies with
               'psd<P
sd=mulcpx(complex(kc),tx) 'remove frequency
               'components from x
td=real( invfft(sd) )    'convert back to time domain
col(2)=td          'place results in worksheet
```

To calculate and graph the smoothing of a set of data using a gain filter, you can either use the provided sample data and graph, or begin a new notebook, enter your own data, and create your own graph using the data.

1. To use the sample worksheet and graph, open the Gain Filter Smoothing worksheet and graph by double-clicking the graph page icon in the Gain Filter Smoothing section of the Transform Examples notebook. Data appears in columns 1 through 3 of the worksheet, and two graphs showing plots, and one

blank graph appear on the graph page. Column 1 contains the Y data for the signal plus noise, column 2 contains the X data and column 3 contains the Y data for the power spectral density graph. The top graph plots the signal plus the noise distortion; the middle graph plots the power spectral density.

2. To use your own data, place your data in column 1. If your data is in a different column, specify the new column after you open the GAINFILT.XFM transform file. If necessary, specify a new column for the results.

3. Press F10 to open the User-Defined Transform dialog box, then click the Open button, and open GAINFILT.XFM transform file in the XFMS directory. The Gain Filter transform appears in the edit window.

Σ   To use this transform, make sure Insert mode is turned off.

4. Click Run. The results are placed in column 5 unless you specified a different column in the transform.

5. If you opened the Gain Filter Smoothing graph, view the graph page. Three graphs appear. The top graph plots the signal plus the noise distortion using a Line Plot with a Simple Straight line style and Single Y data format, plotting column 1 as the Y data for the signal plus noise. The middle graph plots the power spectral density using a Line Plot with a Simple Straight Line style and XY Pairs data format, plotting column 2 as the X data and column 3 as the Y data for the power spectral density graph. The lower graph is a plot of the gain filtered signal, using a Line Plot with a Simple Straight Line style, and single Y data format from column 5.

6. To plot your own data using SigmaPlot, choose the Graph menu Create Graph command, or select the Graph Wizard from the toolbar. Create two graphs. Plot the signal plus the noise distortion using a Line Plot with a Simple Straight line style and Single Y data format, plotting column 1 as the Y data for the signal plus noise. Plot the gain filtered signal using a Line Plot with a Simple Straight Line style, and single Y data format from column 5.

For more information on how to create graphs in SigmaPlot, see the SigmaPlot *User's Manual.*

Figure 6–8
Gain Filter
Smoothing Graph

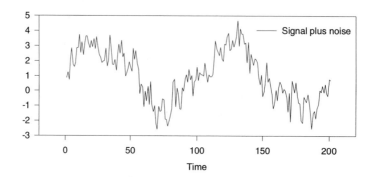

The top graph shows the signal plus noise distortion. The middle graph shows the power spectral density of the signal plus noise distortion. The lower graph shows the gain filter smoothed data.

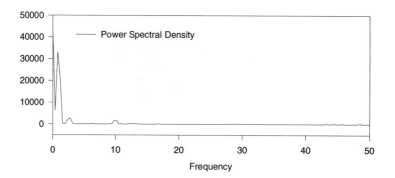

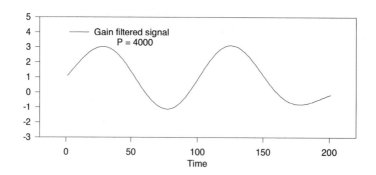

**Gain Filter Transform
(GAINFILT.XFM)**

```
'   Gain Filtering
'      This transform filters data by removing
'      frequency components with power spectral density
'      magnitude less than a specified value
'   Input
```

```
ci = 1     ' input data column number
co = 5     ' output column number
P = 4000    ' psd threshold

' Program
x=col(ci)
n=size(x)
tx = fft(x)            ' compute fft
md = real(tx)^2 + img(tx)^2  ' compute psd
kc = if(md > P,1,0)   ' find frequencies with psd < P
sd = mulcpx(complex(kc),tx)   ' remove frequency
                ' components from x
td = real( invfft(sd) ) 'convert back to the time
              ' domain
nx=size(tx)/2      ' remove padded channels
ru=if(mod(n,2)>0, (nx-n+1)/2, (nx-n+2)/2)
rl=if(mod(n,2)>0, nx-ru, nx-ru+1)

' Output
col(co) = td[data(ru,rl)] ' place results in worksheet
```

Frequency Plot

This transform example creates a frequency plot showing the frequency of the occurrence of data in the Y direction. Data is grouped in specified intervals, then horizontally plotted for a specific Y value. Parameters can be set to display symbols that are displaced a specific distance from each other or that touch or overlap. You can also plot the mean value of each data interval. This transform example shows overlapping symbols which give the impression of data mass.

To calculate and graph the frequency of the occurrence of a set of data, you can either use the provided sample data and graph, or begin a new notebook, enter your own data and create your own graph using the data.

1. To use the sample worksheet and graph, open the Frequency Plot worksheet and graph by double-clicking the graph page icon in the Frequency Plot section of the Transform Examples notebook. Data appears in columns 1 through 3 of the worksheet, and an empty graph appears on the graph page.

2. To use your own data, place your data in columns 1 through 3. You can put data in as many or as few columns as desired, but if you use the sample transform you must change the X locations of the Y values in the second line under the Input heading in the transform file to reflect the number of data columns you are using. If your data is in other columns or more than three columns, specify the new columns after you open the FREQPLOT.XFM transform file.

Enter the tick labels for the X axis in a separate column, and specify tick labels from a column using the Tick Labels Type drop-down list in the Tick Labels panel in Graph Properties Axis tab.

3. Press F10 to open the User-Defined Transform dialog box, then click the Open button, and open the FREQPLOT.XFM transform file in the XFMS directory. The Frequency Plot transform appears in the edit window.

4. Click Run. The results are placed starting one column over from the original data.

5. If you opened the sample Frequency Plot graph, view the graph page. A Scatter Plot appears plotting columns 5 and 6, 7 and 8, and 9 and 10 as three separate XY Pair plots. The lines passing through each data interval is a fourth Line Plot with a Simple Straight Line style plotting columns 11 and 12 as an XY pair, representing the mean value of each data interval. The X axis tick marks are generated by the transform. The axis labels are taken from column 13.

Figure 6–9
Frequency Plot Graph

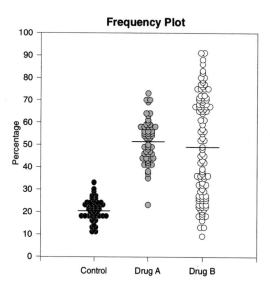

6. To create your own graph using SigmaPlot, make a graph with three Scatter Plots with Simple Scatter styles. Plot each consecutive result column pair as XY pair scatter plots. If the mean line option is active in the transform, plot the last consecutive result column pair as a XY pair Line Plot with Simple Straight Line style. Use labels typed into a worksheet column as the X axis tick labels.

For more information on how to create graphs in SigmaPlot, see the SigmaPlot *User's Manual.*

Frequency Plot Transform (FREQPLOT.XFM)

```
'        *********Frequency Plot********
'        *********FREQPLOT.XFM*********
'            **7-26-95**
```

'This transform creates frequency plots and mean bars 'of multiple y data columns

'It uses data in the first columns of the worksheet and 'creates column pairs for graphing

```
'        ********Procedure - Data Entry*******
```

'1. TURN INSERT OFF
'2. Enter y data groups into columns in the worksheet 'starting with column 1
'3. Select a symbol diameter d (try 0.05 to 0.10 in)
'4. Specify the x locations for the groups (1,2,3,... 'are typically used since ticks are usually labeled)
'Important! Make sure that the number of numbers in 'x={1,2,...} equals the number of y data columns
'5. Enter the width of your graph wg in inches (double 'click on graph to determine its width)
'6. Enter the x range of your graph (usually 1 + number 'of groups)
'7. Enter the vertical data interval w into which data 'points will be grouped
'8. Enter the first vertical data interval start value '(e.g., 0 if the vertical range is 0 to 100)
'9. Enter the horizontal distance fx between symbols '(try 0.05, use negative value for overlap effect)
'10. Specify ml=1 if you want mean lines computed and ' specify mean line width eml
'11. Specify intvl=1, 2 or 3 to place the y data at the 'bottom, center or top of the vertical data interval
```
'    ********Procedure - Graph********
```

'1. Create x,y scatter plots for the column pairs
'2. If mean lines are computed create an x,y line plot ' with no symbols from the last two columns generated

```
'        ********Input*******
```

```
d =.08'size (diameter) of symbol (in)
x={1,2,3}'x locations for groups of y
'values(typically 1,2,3, etc.)
wg=5'width of graph (in)
wd=4'x range of graph (x maximum minus 'x mininum)
fx=0.05'horizontal distance between
'symbols (fraction of symbol
```

```
'diameter)
w=1'vertical data interval (y axis
'units)
ys=0'first vertical data interval
'start value (y axis units)
intvl=3'specifies y display position in w
          'interval(1=bottom,2=center,3=top)
ml=1'include mean lines (0=no, 1=yes)
eml=.6'width of mean line (x axis units)

'       *********Program**********

cy=1'first y group column number
colfi=size(x)
e=1e-18
wx=(1+fx)*d*wd/wg'horizontal distance between
'symbol centers (user units)
ypos=if(intvl=1,w,if(intvl=2,w/2,0))  'y display
          'position
for j = 1 to colfi do   'multiple column loop
coly=col(cy+j-1)-e
buckets=data(ys,max(coly)+w+e,w)
h=histogram(coly,buckets) 'histogram of data
h0=if(h>0,h)    'histogram with zero values
    'excluded
buckets0=if(h>0,buckets)  'corresponding bucket values
hs0=sum(h0)
col(colfi+2*j+1)=lookup(data(1,total(h0)),sum(h0),buckets0)-
ypos    'y values
tem=lookup(data(1,total(h0)),sum(h0),h0)
col(colfi+2*j)=x[j]+ wx*(mod(data(1,size(tem)),tem)-(tem-1)/
2)'x values
col(3*colfi+2,3*j-2,3*j)=if(ml>0,{x[j]-eml/2,x[j]+eml/2,0/
0})'x values for mean lines
col(3*colfi+3,3*j-2,3*j)=if(ml>0,{1,1,0/0} *mean(col(j)))'y
values for mean lines
end for
```

Gaussian Cumulative Distribution from the Error Function

Rational approximations can be used to compute many special functions. This transform demonstrates a polynomial approximation for the error function. The error function is then used to generate the Gaussian cumulative distribution function. The absolute maximum error for the error function approximation is less than 2.5×10^{-5} (M. Abramowitz and L.A. Stegun, *Handbook of Mathematical Functions*, p. 299).

To calculate and graph the Gaussian cumulative distribution for given X values, you can either use the provided sample data and graph or begin a new notebook, enter your own data and create your own graph using the data.

1. To use the sample worksheet and graph, open the Gaussian worksheet and graph by double-clicking the graph page icon in the Gaussian section of the Transform Examples notebook. Data appears in column 1 of the worksheet and two empty graphs appear on the graph page.

2. To use your own data, place the X data in column 1. If your data has been placed in another column, you can specify the column after you open the GAUSDIST.XFM transform file.

3. Press F10 to open the User-Defined Transform dialog box, then click the Open button, and open the GAUSDIST.XFM transform file in the XFMS directory. The Gaussian Cumulative transform appears in the edit window.

4. Click Run. The results are placed in column 2, or in the column specified by the res variable.

5. If you opened the sample Gaussian graph, view the graph page. A Line Plot appears with a spline curve in the first graph with column 1 as the X data versus column 2 as the distribution (Y) data (see Figure 6–10 on page 110).

6. To create your own graph using SigmaPlot, make a Line Plot graph with a Simple Spline Curve. The spline curve plots column 1 as the X data versus column 2 as the distribution (Y) data (see Figure 6–10 on page 110).

   For more information on how to create graphs in SigmaPlot, see the SigmaPlot *User's Manual*.

Gaussian Cumulative Distribution on a Probability Scale

The probability scale is the inverse of the Gaussian cumulative distribution function. When a Gaussian cumulative distribution function is graphed using the probability scale, the result is a straight line.

1. If you opened the sample Gaussian graph, view the graph page. A straight line plot appears in the second graph plotting the distribution data in column 3 along a probability scale.

2. To create your own graph using SigmaPlot, create a Line Plot with a Simple Straight Line using column 1 as your X data and column 3 as your Y data, and set the Y axis scale to Probability.

   For information on how to create graphs in SigmaPlot, see the SigmaPlot *User's Manual*.

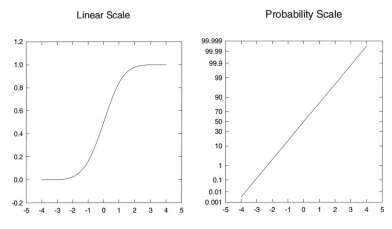

Figure 6–10
Gaussian Cumulative
Distribution Graphs

Gaussian Cumulative Distribution Function

Linear Scale   Probability Scale

Gaussian Distribution (GAUSDIST.XFM)

```
'*** Gaussian Cumulative Distribution Function ***
'************** (C.D.F.) Transform **************
' This transform takes x data and returns the
' results of a Gaussian Cumulative Distribution
' function
' Place your x data in x_col or change the column
' number to suit your data. Results are placed in
' column res
x_col=1      'column for x data
res=2        'column for Gaussian Cumulative
         'Distribution values
x=col(x_col)  'define x values
'*** CALCULATE POLYNOMIAL APPROXIMATION TO THE ***
'**************** ERROR FUNCTION ****************
' You can place the functions erf(x) and terf(x)
' in the Transform Library to create user-defined
' functions for the error function.
erf(x)=1-(.3480242*terf(x)-.0958798*terf(x)^2
    +.7478556*terf(x)^3)*exp(-x^2)
terf(x)=1/(1+.47047*x)
erf1(x)=if(x<0,-erf(-x),erf(x))
'** CALCULATE GAUSSIAN CUMULATIVE DISTRIBUTION **
'******** AND PLACE RESULTS IN WORKSHEET ********
P(x)=(erf1(x/sqrt(2))+1)/2 'Gaussian C.D.F.

col(res)=P(x)
col(res+1)=col(res)*100
```

Histogram with Gaussian Distribution

This transform calculates histogram data for a normally distributed sample, then uses the sample mean and standard deviation of the histogram to compute and graph a Gaussian distribution for the histogram data.

The Histogram Gaussian transform uses examples of the following functions:

➤ gaussian

➤ histogram

➤ size

➤ [...] (array reference)

To calculate and graph a histogram and Gaussian curve for a normally distributed sample, you can either use the provided sample data and graph or begin a new notebook, enter your own data, and create your own graph using the data.

To use the sample worksheet and graph:

1. Open the Histogram Gaussian worksheet and graph by double-clicking the graph page icon in the Histogram Gaussian section of the Transform Examples notebook. The Histogram worksheet with data in column 1 and an empty graph page appears.

    The data in the Histogram Gaussian worksheet was generated using the transform:

    col(1) = gaussian(100,0,325,2)

To use your own data:

1. Place the sample in column 1 of the worksheet. If your data has been placed in another column, you can specify this column after you open the HIST-GAUS.XFM transform file. You can enter the data into an existing or new worksheet.

2. Press F10 to open the User-Defined Transform dialog box, then click the Open button, and open the HISTGAUS.XFM transform file in the XFMS directory. The Histogram with Gaussian Distribution transform appears in the edit window.

3. Click Run. The results are placed in columns 2 through 5 of the worksheet, or in the columns specified by the res variable.

4. If you opened the Histogram Gaussian graph, view the graph page. A histogram appears using column 2 as X data versus column 3 as the Y data. The curve plots the Gaussian distribution using column 4 as X data versus column 5 as the Y data.

5. To create your own graph using SigmaPlot, create a simple vertical bar chart and

set the bar widths as wide as possible. Add the Gaussian curve to the graph by creating another plot using the data in column 4 as the X data and the data in column 5 as the Y data.

Figure 6–11
The Histogram
Gaussian Graph

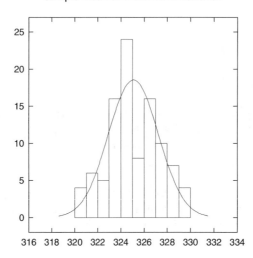

Gaussian Distribution Using the
Sample Mean and Standard Deviation

Histogram with
Gaussian Distribution
Transform
(HISTGAUS.XFM)

```
'******* Transform for a Histogram with a *******
'****** Superimposed Gaussian Distribution ******
' This transform can be used to create histogram
' values for a sample with a normal distribution
' and the data for a smooth Gaussian curve for the
' histogram
' Place your normally distributed sample data in
' x_col or change the column number to suit your
' data. Results are placed in columns res through
' rc+3.
x_col=1    'column number for sample data
res=2      'first results column
'Set histogram range:
min=318    'left limit of histogram
max=334    'right limit of histogram
interval=1 'histogram interval
'define source data:
x=col(x_col)
'*********** GENERATE HISTOGRAM DATA ***********
historange = data(min,max,interval)
h=histogram(x,historange)
```

```
int2=interval/2
y=h[data(2,size(h)-1)]
'***** PLACE HISTOGRAM XY DATA IN WORKSHEET *****
'bar positions (x values):
col(res)=historange[data(1,size(historange)-1)]
 +int2
'bar heights (y values):
col(res+1)= y
'*** GENERATE GAUSSIAN DISTRIBUTION CURVE DATA ***
pi=3.1415926
m=mean(x)
s=stddev(x)
x1=data(m-3*s,m+3*s,6*s/20)
y1=exp(-((x1-m)/s)^2/2)/(sqrt(2*pi)*s)
'**** PLACE GAUSSIAN CURVE DATA IN WORKSHEET *****
col(res+2)= x1
col(res+3)= y1*interval*total(y)
```

Linear Regression with Confidence and Prediction Intervals

This transform computes the linear regression and upper and lower confidence and prediction limits for X and Y columns of equal length. A rational polynomial approximation is used to compute the *t* values used for these confidence limits.

Figure 6–12 displays the sample Linear Regression graph with the results of the LINREGR.XFM transform plotted.

The LINREGR.XFM transform contains examples of these two functions:

➤   min
➤   max

To calculate and graph a linear regression and confidence and prediction limits for XY data points, you can either use the provided sample data and graph or begin a new notebook, enter your own data, and create your own graph using the data.

1.   To use the provided sample data and graph, open the Linear Regression worksheet and graph by double-clicking the graph page icon in the Linear Regression section of the Transform Examples notebook. The worksheet appears with data in columns 1 and 2. The graph page appears with a scatter graph plotting the original data in columns 1 and 2.

2.   To use your own data, place the X data in column 1 and the Y data in column 2. If your data has been placed in other columns, you can specify these columns after you open the LINREGR.XFM transform file. You can enter data into an existing or a new worksheet.

3.   Press F10 to open the User-Defined Transform dialog box, then click the Open button, and open the LINREGR.XFM transform in the XFMS directory. The

Linear Regression transform appears in the edit window. If necessary, change the x_col, y_col, and res variables to the correct column numbers (this is not necessary for the example Linear Regression worksheet data).

4. Change the Z variable to reflect the desired confidence level (this is not necessary for the example Linear Regression worksheet data).

5. Click Run. The results are placed in columns 3 through 8, or in the columns specified by the res variable.

6. If you opened the Linear Regression graph, view the graph page. The original data in columns 1 and 2 is plotted as a scatter plot. The regression is plotted as a solid line plot using column 3 as the X data versus column 4 as the Y data, the confidence limits are plotted as dashed lines using column 3 as a single X column versus columns 7 and 8 as many Y columns, and the prediction limits are plotted as dotted lines using column 3 as a single X column versus columns 7 and 8 as many Y columns.

7. To create your own graph in SigmaPlot, create a Scatter Plot with a Simple Regression, plotting column 1 against column 2 as the symbols and using column 3 plotted against column 4 as the regression. Add confidence and prediction intervals using column 3 as the X column and columns 7 and 8 as the Y columns.

For more information on creating graphs in SigmaPlot, see the SigmaPlot *User's Manual.*

Figure 6–12
Linear Regression Graph

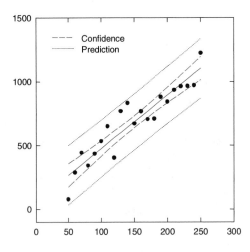

95% Confidence and Prediction Intervals

Linear Regression Transform (LINREGR.XFM)

```
'*** Transform to Compute a Linear Regression ***
'**** with Confidence & Prediction Intervals ****

' Place your x data in x_col and y data in y_col or
' change the column numbers to suit your data.
' Results are placed in columns res through res+5.

x_col=1        'column number for x data
y_col=2        'column number for y data
res=3          'first results column

x=col(x_col)     'Define x values
y=col(y_col)     'Define y values

'Define z value for 95% confidence interval
'for 99% confidence interval, use z=2.576

z=1.96         'z for 95% confidence
'z=2.576        'z for 99% confidence
;********* DEFINE REGRESSION PARAMETERS **********
n=size(x)            'number of data points
v=n-2                'n must be > 2
xbar=mean(x)          'mean of x
denom=total((x-xbar)^2)   'sum of sqs about mean
alpha=total(x^2)/(n*denom) '1,1 coeff of (X'X)^-1
beta=-xbar/denom        '1,2 coeff of (X'X)^-1
delta=1/denom           '2,2 coeff of (X'X)^-1
r1=total(y)            '1st row of X'Y
r2=total(x*y)          '2nd row of X'Y
b0=alpha*r1+beta*r2      'intercept parameter
b1=beta*r1+delta*r2      'slope parameter
'*** CALCULATE REGRESSION AND CONFIDENCE DATA ***
'Regression data
xreg=data(min(x),max(x),(max(x)-min(x))/20)
yreg=b0+b1*xreg

'Compute t value
t123=z+(z^3+z)/(4*v)+(5*z^5+16*z^3+3*z)/(96*v^2)
t4=(3*z^7+19*z^5+17*z^3-15*z)/(384*v^3)
t5=79*z^9+776*z^7+1482*z^5-1920*z^3-945*z
t=t123+t4+t5/(92160*v^4)

'Estimate of sigma
s=sqrt(total(((y-(b0+b1*x))^2))/v)
'Confidence Limit data
term=alpha+2*beta*xreg+delta*xreg^2
conf_lim=sqrt(term)
```

```
up_conf=yreg+t*s*conf_lim      'upper limit
low_conf=yreg-t*s*conf_lim     'lower limit
'Prediction Intervals data
pred_lim=sqrt(1+term)
up_pred=yreg+t*s*pred_lim 'upper prediction limit
low_pred=yreg-t*s*pred_lim 'lower prediction limit


'******* PLACE REGRESSION AND CONFIDENCE *********
'************* DATA IN WORKSHEET ***************
'Regression
col(res)=xreg       'x values of regression line
col(res+1)=yreg      'y values of regression line

'Confidence Interval
col(res+2)=up_conf  'upper confidence limit
col(res+3)=low_conf  'lower confidence limit

'Prediction
col(res+4)=up_pred  'upper prediction limit
col(res+5)=low_pred  'lower prediction limit
```

Low Pass Filter

This transform is a smoothing filter which produces a data sequence with reduced high frequency components. The resulting data can be graphed using the original X data.

To calculate and graph a data sequence with reduced high frequency components, you can either use the provided sample data and graph or begin a new notebook, enter your own data, and create your own graph using the data.

1. To use the provided sample data and graph, double-click the Low Pass Filter graph page icon in the Low Pass Filter section of the Transform Examples notebook. The worksheet appears with data in columns 1 and 2. The graph page appears with two graphs. The first is a line graph plotting the raw data in columns 1 and 2 (see Figure 6–11 on page 112). The second graph is empty.

2. To use your own data, place your Y data (amplitude) in column 2 of the worksheet, and the X data (time) in column 1. If your data is in other columns, you can specify these columns after you open the LOWPFILT.XFM file. You can enter your data in an existing or new worksheet.

3. Press F10 to open the User-Defined Transform dialog box, then click the Open button, and open the LOWPFILT.XFM transform file in the XFMS directory. The Low Pass Filter transform appears in the edit window.

4. Set the sampling interval **dt** (the time interval between data points) and the half power point **fc** values. The half power point is the frequency at which the

squared magnitude of the frequency response is reduced by half of its magnitude at zero frequency.

5. If necessary, change the **cy1** source column value and **cy2** filtered data results to the correct column numbers.

6. Click Run to run the transform. Filtered data appears in column 3 in the worksheet, or in the worksheet column you specified in the transform.

7. If you opened the Low Pass Filter graph, view the graph page. The second graph appears as a line graph plotting the smoothed data in columns 1 and 3.

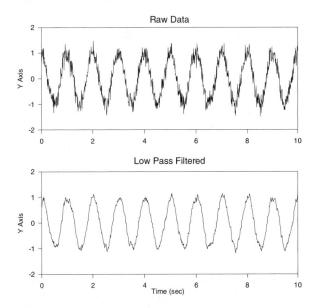

Figure 6–13
Low Pass Filter Graph
Plotting Raw Data
and Filtered Data

8. To create your own graphs in SigmaPlot, create the first graph as a Line Plot with a Simple Spline Curve using the raw data in columns 1 and 2 as the X and Y data. Make the second Line Plot graph with a Simple Spline Curve using the data in column 1 as the X data and the smoothed data in column 3 as the Y data.

    For more information on creating graphs in SigmaPlot, see the SigmaPlot *User's Manual*.

Low Pass Filter Transform

```
'**** First Order Low Pass Recursive Filter ****
'This filter will smooth data by reducing
'frequency components above the half power point.
'It generates the filtered output y(i) from the
'data x(i).

'y(i) = a y(i-1) + (1 - a) x(i)
```

```
'where a is computed from the specified half power
'point fc.

'        ********** Input **********

dt = .01   'sampling interval (sec)
fc = 5     'half power point of filter (Hz)
cy1 = 2    'column number for input data

'        ********** RESULTS **********

cy2 = 3    'column number for filtered output

'        ********** Program **********

pi=arccos(-1)
cos2pft = cos(2*pi*fc*dt)
a = 2-cos2pft - sqrt(cos2pft^2-4*cos2pft+3)
               'filter coefficient
cell(cy,1)=cell(cx,1)    'recursive filter
for i=2 to size(col(cx)) do
 cell(cy,i)=a*cell(cy,i-1)+(1-a)*cell(cx,i)
end for
```

Lowess Smoothing

Smoothing is used to elicit trends from noisy data. Lowess smoothing produces smooth curves under a variety of conditions[1]. "Lowess" means locally weighted regression. Each point along the smooth curve is obtained from a regression of data points close to the curve point with the closest points more heavily weighted.

The y value of the data point is replaced by the y value on the regression line. The amount of smoothing, which affects the number of points in the regression, is specified by the user with the parameter f. This parameter is the fraction of the total number of points that is used in each regression. If there are 50 points along the smooth curve with $f = 0.2$ then 50 weighted regressions are performed and each regression is performed using 10 points.

An example of the use of lowess smoothing for the U.S. wheat production from 1872 to 1958 is shown in the figures below. The smoothing parameter f was chosen to be

1. *Visualizing Data*, William S. Cleveland

0.2 since this produced a good tradeoff between noisy undersmoothing and oversmoothing which misses some of the peak-and-valley details in the data.

Figure 6–14
U.S. Wheat data and the
lowess smoothed curve (f =
0.2). Notice the definite
decreased production during
World War II.

Lowess Smoothed U.S. Wheat Production

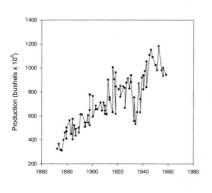

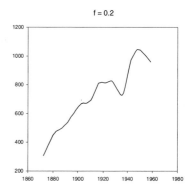

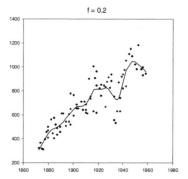

1. **To use the provided sample data and graph**, open the Lowess Smoothing worksheet and graph in the Lowess Smoothing section of the Transform Examples notebook. The worksheet appears with data in columns 1, 2, and 3.

2. To use your own data, enter the XY data for your curve in columns 1 and 2, respectively. If your data has been placed in other columns, you can specify these columns after you open the **LOWESS.XFM** transform file. Enter data into an existing or a new worksheet.

3. Press F10 to open the User-Defined Transform dialog box, then click the Open button, and open the LOWESS.XFM transform file in the XFMS directory. The Lowess transform appears in the edit window.

4. Click Run. The results are placed in column 3 of the worksheet, or in the column specified by the ouput variable.

5. If you opened the Lowess Smoothing graph, view the graph page. The smoothed curve is plotted on the second graph and both the orginal and smoothed data are plotted on the third.

If you want to plot your own results, create a line plot of column 1 versus column 3.

For more information of creating graphs, see the SigmaPlot *User's Manual.*

LOWESS.XFM

```
'******Lowess Smoothing Example******
x=col(1)
y=col(2)
f=0.2

'******Results******
col(3)=output

'******Program******
output=lowess(x,y,f)
```

Normalized Histogram

This simple transform creates a histogram normalized to unit area. The resulting data can be graphed as a bar chart. Histogram bar locations are shifted to be placed over the histogram box locations. The resulting bar chart is an approximation to a probability density function (see Figure 6–16 on page 123).

To calculate and graph a normalized histogram sample, you can either use the provided sample data and graph or begin a new notebook, enter your own data, and create your own graph using the data.

1. To use the provided sample data and graph, open the Normalized Histogram worksheet and graph in the Normalized Histogram and Graph section of the Transform Examples notebook. The worksheet appears with data in column 1. The data is made up of exponentially distributed random numbers generated with the transform:

   $x = random(200,1,1.e-10,1)col(1) = -ln(x)$

   The graph page appears with an empty graph.

2. To use your own data, place your data in column 1 of the worksheet. If your data has been placed in another column, you can specify this column after you open the NORMHIST.XFM transform file. You can enter data into an existing or new worksheet.

3. Press F10 to open the User-Defined Transform dialog box, then click the Open button, and open the NORMHIST.XFM transform file in the XFMS directory. The Normalized Histogram transform appears in the edit window.

4. Click Run. The results are placed in columns 2 and 3 of the worksheet, or in the columns specified by the res variable.

5. If you opened the Normalized Histogram graph, view the graph page. A histogram appears using column 2 as X data versus column 3 as the Y data.

6. To create your own graph in SigmaPlot, create a Vertical Bar chart with simple bars, then set the bar widths as wide as possible.

For more information of creating bar charts and setting bar widths, see the SigmaPlot *User's Manual.*

Figure 6–15
Normalized
Histogram Graph

Normalized Histogram of Exponential Random Numbers

n=200

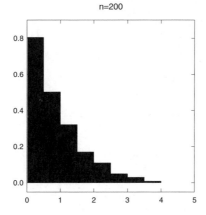

Normalized
Histogram Transform
(NORMHIST.XFM)

```
'** Transform to Generate Normalized Histogram **

' This transform normalizes a bar chart to unit
' area. Box locations are shifted since bars are
' drawn about their center points

' Place your sample data in x_col or change the
' column number to suit your data. Results are
' placed in columns res through res+1.

x_col=1            ' sample data column
res=2              'first results column

'Set histogram box width and upper limit
limit=5.5              'upper limit of last box
delta=0.5              'histogram box width

x=col(x_col)
```

```
'**** CALCULATE AND NORMALIZE HISTOGRAM DATA ****

r=data(delta,limit,delta) 'histogram boxes
h=histogram(x,r)        'create histogram
h1=h[data(1,size(h)-1)]  'remove last box

'shift bar center locations
x1=r-delta/2           'histogram x values

'normalize histogram
y1=h1/(total(h1)*delta)  'histogram Y values

'* PLACE NORMALIZED HISTOGRAM DATA IN WORKSHEET *

col(res)=x1
col(res+1)=y1
```

Shading Beneath Line Plot Curves

These are a pair of transforms that use two different methods to draw colors or hatches below a curve.

SHADE_1.XFM uses bar chart fills or colors to fill the area below a curve. This method must be used if you want to fill with a color; however, you can only shade to an axis, and you can only use the default Windows fill patterns.

SHADE_2.XFM uses line plots to fill below curves. This transform can also be used to draw fill lines between two curves.

Shading Below a Curve with Color

To use this transform to create a shade under a curve, you can either use the provided sample data and graph or begin a new notebook, enter your own data, and create your own graph using the data.

1. To use the provided sample data and graph, open the Shade 1 worksheet and graph by double-clicking the graph page icon in the Shade 1 section of the Transform Examples notebook. The worksheet appears with data in columns 1 and 2. The graph page appears with a line graph plotting column 1 against column 2.

2. To use your own data, enter the XY data for your curve in columns 1 and 2, respectively. If your data has been placed in other columns, you can specify these columns after you open the **SHADE_1.XFM** transform file. Enter data into an existing or a new worksheet.

3. Create a line graph with two curves using your own data by creating a Line Plot with a Simple Straight Line curve plotting the column 1 data against the column 2 data.

4. Press F10 to open the User-Defined Transform dialog box, then open the SHADE_1.XFM transform file. The Shade 1 transform appears in the edit window. If necessary, change the source column numbers **x_data** and **y_data** to the correct column numbers.

5. Click Run. The data for the bar chart is placed in columns 3 and 4, or whatever columns were specified.

6. If you opened the sample Shade 1 graph, view the graph page. The graph automatically appears plotting the curve of the original data and the data representing the shade under the curve.

7. If you created your own graph (see step 3) and you want to use SigmaPlot to plot the shade under the curve, add the shade under the curve by creating a Simple Vertical Bar Chart that plots columns 3 and 4, setting the bar width to maximum and the bar fill to either no pattern and the same color fill and edge as the line, or with a default Windows hatch pattern and fill and edge colors of none.

For more information on creating graphs in SigmaPlot, see the SigmaPlot *User's Manual.*

Figure 6–16
The Shade 1 Graph

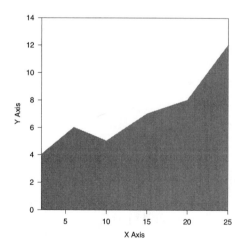

Shade Beneath Curve
Transform
(SHADE_1.XFM)

```
'    ****** Shading to an Axis ******
'This transform uses vertical bars to create a
'fill between a curve and an axis. The x data
'MUST be sorted in increasing or decreasing order.
```

```
'Apply this transform to your x,y data column pair.

'     ************ Input ************
x_data = col(1) 'column for x data
y_data = col(2) 'column for y data
density = 200

' The density determines how many bars are used
'to create the fill under the curve. The larger
'the density, the more bars are used, and the
'longer the graph takes to draw or print. For
'relatively "flat" curves, try using a smaller
'value for the density (like about 150). For
'sharply peaked curves, it may be necessary to
'increase the value of the density (to about 350).

'     ********** Output **********
x_result = 3  ' column for x patterned bar fill
y_result = 4  ' column for y patterned bar fill

'     ************ Program ************
dmax = max(x_data)
dmin = min(x_data)
dx = (dmax-dmin)/density
x = data(dmin,dmax,dx)
y = interpolate(x_data,y_data,x)
col(x_result) = x
col(y_result) = y
```

Shading Between Curves

To use this transform to create a shade pattern between two curves, you can either use the provided sample data and graph, or begin a new notebook, enter your own data, and create your own graph using the data.

1. To use the provided sample data and graph, open the Shade 2 worksheet and graph by double-clicking the graph page icon in the Shade 2 section of the Transform Examples notebook. The worksheet appears with data in columns 1 through 4. The graph page appears with a line and scatter graph with two curves plotting column 1 against column 2 and column 3 against column 4.

2. To use your own data, enter the XY data for the first curve in columns one and two, and the XY data for the second curve in columns three and four, respectively. The X data for both curves must be in strictly increasing order. If your data has been placed in other columns, you can specify these columns after you open the **SHADE_2.XFM** transform file. You can enter data into an existing or a new worksheet.

3. To use your own data to create a graph, make a Line Plot with Multiple Straight Lines plotting the column 1 data against column 2 data for the first curve, and column 3 data against column 4 data as the second curve.

4. Press F10 to open the User-Defined Transform dialog box, then click the Open button, and open the SHADE_2.XFM transform in the XFMS directory. If necessary, change the source column numbers to the correct column numbers.

5. Set the fill density to use. For a solid color between the curves, use a large density, about 500. For a nicely spaced vertical fill, try a density of 50.

6. Click Run. The data for the shade lines is placed in columns 5 and 6, or whatever columns were selected.

7. If you opened the Shade 2 graph, view the graph page. The graph automatically appears with the new plot filling the space between the curves in the original plot

8. If you created your own graph (see step 3) and you want to use SigmaPlot to plot the shade between the curves, add a new Line and Scatter Plot with Multiple Straight Lines to the graph using columns 5 and 6 for the X and Y data, and if necessary, turn symbols off. The new plot appears as shade between the curves of the original plot.

For more information on creating graphs in SigmaPlot, see the SigmaPlot *User's Manual*.

Figure 6–17
The Shade 2 Graph

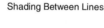

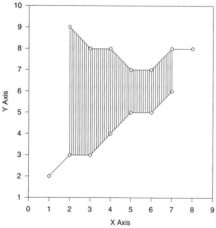

Shade Between Curves Transform (SHADE_2.XFM)

```
'    ****** Shading Between Curves ******
'This transform fills the area between two x-y
'line curves with vertical lines. The x data
'for both curves MUST be in strictly increasing
'order.
'Create a new line and scatter plot, select
''result_x' and 'result_y' columns for the x
'and y axes.; Turn symbols off in the Symbols panel of
'the Plot tab in the Graph Properties dialog box.
'For a solid color between the curves, use a
'large density, about 500. For a nicely spaced
'vertical fill, try a density of 50.

'    *********** Input ************

cx1=1      'column for first curve x data
cy1=2      'column for first curve y data
cx2=3      'column for second curve x data
cy2=4      'column for second curve y data
density = 50   'line density of fill

'    ********** RESULTS ************

result_x = 5   'column for x fill results
result_y = 6   'column for y fill results

'    ********** Program ***********
X1 = col(cx1)   'x data for first curve
Y1 = col(cy1)   'y data for first curve
X2 = col(cx2)   'x data for second curve
Y2 = col(cy2)   'y data for second curve
x1_min = min(X1)
x2_min = min(X2)
x1_max = max(X1)
x2_max = max(X2)

'Take the largest x_min and the smallest x_max
x_min = if(x1_min < x2_min,x2_min,x1_min)
x_max = if(x1_max > x2_max,x2_max,x1_max)

dx = abs(x_max - x_min)/density
x = data(x_min,x_max,dx)
y1 = interpolate(X1,Y1,x)
y2 = interpolate(X2,Y2,x)
```

```
a = interpolate(x,x,x)

for i = 1 to size(a) do
cell(result_x,3*i-2) =a[i]
cell(result_x,3*i-1) =a[i]
cell(result_x,3*i) = 0/0

cell(result_y,3*i-2) = y1[i]
cell(result_y,3*i-1) = y2[i]
cell(result_y,3*i) = 0/0
end for
```

Smooth Color Transition Transform

This transform example creates a smooth color transition corresponding to the changes across a range of values. The transform places color cells in a worksheet column that change from a specified start color to a specified end color, each color cell incrementing an equivalent shade for each data value in the range. This transform example shows how the color transform can be set to display a "cool" (blue) color that corresponds to small residuals, and a "hot" (red) color that corresponds to large residuals resulting from a nonlinear regression. Since residuals vary positively and negatively about zero, the absolute values for the residuals are used in the transform.

Σ   It is unnecessary to sort the data before executing the smooth color transition transform.

To calculate and graph the smooth color transition of a set of data, you can either use the provided sample data and graph, or begin a new notebook, enter your own data, and create your own graph using the data.

1.   To use the sample worksheet and graph, open the Smooth Color Transition worksheet and graph by double-clicking the graph page icon in the Smooth Color Transition section of the Transform Examples notebook. Data appears in columns 1 and 2 of the worksheet, and a scatter graph appears on the graph page.

2.   To use your own data, place your data in columns 1 and 2. For the residuals example, column 2 is the absolute value of the residuals in column 1. To obtain absolute values of your data, use the abs transform function. For example, to obtain the absolute values of the data set in column 1, type the following transform in the User-Defined Transform dialog box:

col(2)=abs(col(1))

If your data is in a different column, specify the new column after you open the RGBCOLOR.XFM transform file.

3. Press F10 to open the User-Defined Transform dialog box, then click the Open button, and open the RGBCOLOR.XFM transform file in the XFMS directory. The Smooth Color Transition transform appears in the edit window.

4. Click Run. The results are placed starting one column over from the original data, or in the column you specified in the transform.

5. If you opened the sample Smooth Color Transition graph, view the graph page. A Scatter Plot appears plotting column 2 as a Simple Scatter plot style using Single Y data format. The symbol colors are obtained by specifying column 3 in the Symbols, Fill Color drop-down list in the Plots panel of the Graph Properties dialog box. The Smooth Color Transition transform applies gradually changing colors to each of the data points. The smaller residual values are colored blue, which gradually changes to red for the larger residuals.

Figure 6–18
Smooth Color Transition
Transform Example Graph

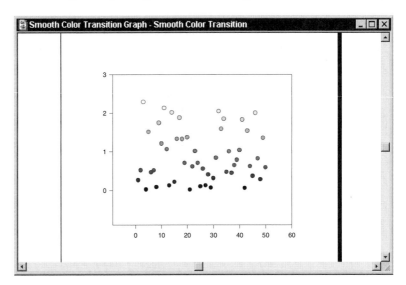

6. To create your own graph using SigmaPlot, make a Scatter Plot graph with a Scatter Plot with Simple Scatter style. Plot the data as Single Y data format. Use the color cells produced by the transform by selecting the corresponding worksheet column from the Symbol Fill Color drop-down list.

For more information on how to create graphs in SigmaPlot, see the SigmaPlot *User's Manual*.

Smooth Color
Transition Transform
(RGBCOLOR.XFM)

```
' Smooth Color Shade Transition from a Data Column
'    This transform creates a column of colors which
'    change smoothly from a user defined initial
'    intensity to a final intensity as the data
'    changes from its minimum value to its maximum
'    value.
```

```
' Input
ci = 2        ' data input column
co = 3        ' color output column
sr =0         ' initial red intensity
sg = 50       ' initial green intensity
sb = 255      ' initial blue intensity
fr = 255      ' final red intensity
fg = 50       ' final green intensity
fb = 0        ' final blue intensity

' Program
d = max(col(ci))-min(col(ci))
range = if( d=0, 1, d)
t = (col(ci) - min(col(ci)))/range
r = (fr-sr)*t+sr
g = (fg-sg)*t+sg
b = (fb-sb)*t+sb

' Output
col(co) = rgbcolor(r,g,b) ' place colors into worksheet
```

Survival (Kaplan-Meier) Curves with Censored Data

This transform creates Kaplan-Meier survival curves with or without censored data. The survival curve may be graphed alone or with the data.

To use the transform, you can either use the provided sample data and graph or begin a new notebook, enter your own data, and create your own graph using the data.

1. To use the sample worksheet and graph, double-click the graph page icon in the Survival section of the Transforms Examples notebook. The Survival worksheet appears with data in columns 1 and 2. The graph page appears with an empty graph. If you open the sample worksheet and graph, skip to step 7.

2. To use your own data, enter survival times in column 1 of the worksheet. Ties (identical survival times) are allowed. You can enter data into an existing or a new worksheet.

3. Enter the censoring identifier in column 2. This identifier should be 1 if the corresponding data point in column 1 is a true response, and 0 if the data is censored.

4. If desired, save the unsorted data by copying the data to two other columns.

5. Select columns 1 and 2, then choose the Transforms menu Sort Selection command. Specify the *key column* in the Sort Selection dialog box as column 1, and

Figure 6–19
The Survival Graph

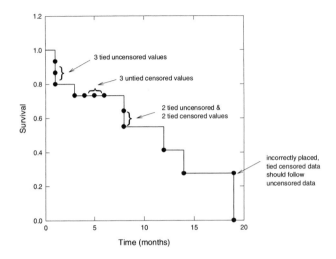

Survival Curve Example with Censored Data

the sort order option as Ascending.

6. Check for any ties between true response and censored data. If any exist, make sure that within the tied data, the censored data follows the true response data.

7. From the worksheet, press F10 to open the User-Defined Transform dialog box, then click the Open button, and open the SURVIVAL.XFM transform in the XFMS directory.

8. Click Run to run the file. The sorted time, cumulative survival probability, and the standard error are placed in columns res, res+1, and res+2, respectively. For graphical purposes a zero, one, and zero have been placed in the first rows of the sorted time, cumulative survival curve probability and standard error columns.

9. If you opened the sample Survival graph, view the page. The Simple Horizontal Step Plot graphs the survival curve data from columns res as the X data versus column res+1 as the Y data and a Scatter Plot graphs the data from the same columns. The first data point of the Scatter Plot at (0,1) is not displayed by selecting rows 2 to end in the Portions of Columns Plotted area of the Data section in the Plots tab of the Graph Properties dialog box. As shown in Figure 6–19, a tied censored data point has been incorrectly placed; it should follow uncensored data.

10. To graph a survival curve using SigmaPlot, create a Line graph with a Simple Horizontal Step Plot graphing column res as the X data versus column res+1 as the Y data. If desired, create an additional Scatter plot, superimposing the survival data using the same columns for X data and Y data. To turn off the symbol drawn at x = 0 and y = 1, select Plot 2 and set Only rows = 2 to end by 1 in the

Plots tab and Data sections of the Graph Properties dialog box.

For more information on creating graphs, see the *SigmaPlot User's Manual.*

Survival Transform (SURVIVAL.XFM)

```
'*Kaplan-Meier Survival Curves with Censored Data*

' This transform calculates cumulative survival
' probabilities and their standard errors
' Enter survival times in column sur_col and a
' censor index in column cen_col (0=censored,
' 1=not), or change the column numbers to suit
' your data. Results are placed in columns res
' and res+1.
' Procedure:
' 1) sort by increasing survival time
' 2) place censored data last if ties
' 3) run this transform
' 4) plot survival data as columns 3 vs 4, as
'    a stepped line shape with symbols
sur_col=1
cen_col=2
res=3
sur=col(sur_col)  'survival data
cen=col(cen_col)  'censored data
'********* CALCULATE CUMULATIVE SURVIVAL *********
mv=0/0         'missing value
i=data(1,size(sur)) 'integers
N=size(sur)       'number of cases
n=N+1
pi=(N-i+1-cen)/(N-i+1)
cs=10^(sum(log(pi)))  'cumulative survival
'Calculate standard error of survival
se=cs*sqrt(sum(cen/((N-i)*(N-i+1))))
'********* PLACE RESULTS IN WORKSHEET *********
col(res)={0,sur}
col(res+1)={1,cs}  'cumulative survival probability
col(res+2)={0,se}  'standard error of survival
```

User-Defined Axis Scale

The USERAXIS.XFM transform is a specific example how to transform data to fit the user-defined axis scale.

$$log\left(log\left(\frac{100}{y} \right) \right)$$

This transform:

➤ transforms the data using the new axis scale

➤ creates Y interval data for the new scale

To use this transform to graph data along a $log\left(log\left(\frac{100}{y}\right)\right)$ Y axis, you can either use the provided sample data and graph, or begin a new notebook, enter your own data, and create your own graph using the data.

1. To use the sample worksheet and graph, double-click the graph page icon in the User Defined Axis Scale section of the Transforms Examples notebook. The User Defined Axis Scale worksheet appears with data in columns 1 through 3. The graph page appears with an empty graph with gridlines.

2. To use your own data, place your original X data in column 1, Y data in column 2, and the Y axis tick interval values in column 3. If your data has been placed in other columns, you can specify these columns after you open the USER-AXIS.XFM file.

3. Press F10 to open the User-Defined Transform dialog box, then open the USERAXIS.XFM transform. If necessary, change the y_col, tick_col, and res variables to the correct column numbers.

4. Click Run. The results are placed in columns 4 and 5, or the columns specified by the res variable.

5. If you opened the User Defined Axis Scale graph, view the page. The graph is already set up to plot the data and grid lines.

6. To plot the transformed Y data using SigmaPlot, plot column 1 as the X values versus column 4 as the Y values.

    To plot the Y axis tick marks, open the Ticks panel under the Axes tab of the Graph Properties dialog box. Select Column 5 from the Major Tick Intervals drop-down list.

    To draw the tick labels, use the Y tick interval data as the tick label source by selecting Column 3 from the Tick Label Type drop-down list in the Tick Labels panel under the Axes tab of the Graph Properties dialog box.

    For more information on creating graphs and modifying tick marks and tick labels, see the SigmaPlot *User's Manual*.

Figure 6–20
User-Defined Axis
Scale Graph

**User-Defined Axis
Scale Transform
(USERAXIS.XFM)**

```
'** Transform for the User Defined Y Axis Scale **
'************* f(y)=log(log(100/y)) *************

' This transform is an example of how to transform
' your data to fit a custom axis scale, and to
' compute the grid line intervals for that scale

' Place your y data in y_col and the y tick mark
' locations in tick_col or change the column
' number to suit your data. Results are placed
' beginning in column res

y_col=2
tick_col=3
res=4

y_data=col(y_col) 'Original y data
'*********** FUNCTION FOR Y AXIS SCALE ***********
f(y)=log(log(100/y))   'Transform for axis scale
'*********** COMPUTE Y TICK INTERVAL DATA ***********
y1=f(col(tick_col))      'y tick intervals
'*** PLACE Y DATA AND Y AXIS GRID IN WORKSHEET ***
col(res)=f(y_data)       'transformed y data
col(res+1)=y1            'y values for y grid
```

Vector Plot

The VECTOR.XFM transform creates a field of vectors (lines with arrow heads) from data which specifies the X and Y position, length, and angle of each vector. The data is entered into four columns. Executing the transform produces six columns of three XY pairs, which describe the arrow body and the upper and lower components of the arrow head.

Other settings are:

➤ the length of the arrow head

➤ the angle in degrees between the arrow head and the arrow body

➤ the length of the vector (if you want to specify it as a constant)

To generate a vector plot, you can either use the provided sample data and graph or begin a new notebook, enter your own data, and create your own graph using the data.

1. To use the sample worksheet and graph, double-click the graph page icon in the Vector section of the Transform Examples notebook. The Vector worksheet appears with data in columns 1 through 4. The graph page appears with an empty graph.

2. To use your own data, enter the vector information into the worksheet. Data must be entered in four column format, with the XY position of the vector starting in the first column, the length of the vectors (which correspond to the axis units), and the angle of the vector, in degrees. The default starting column for this block is column one.

3. Press F10 to open the User-Defined Transforms dialog box, then click the Open button to open the VECTOR.XFM file in the XFMS directory.

4. If necessary, change the starting worksheet column for your vector data block **xc**.

5. If desired, change the default arrowhead length **L** (in axis units) and the **Angle** used by the arrowhead lines. This is the angle between the main line and each arrowhead line.

6. If you want to use vectors of constant length, set the **l** value to the desired length, then uncomment the remaining two lines under the Constant Vector Length heading.

7. Make sure that Radians are selected as the Trigonometric Units (they should be by default.

8. Click Run to run the transform. The transform produces six columns of three XY pairs, which describe the arrow body and the upper and lower components of the arrow head.

9. If you opened the Vector graph, view the page. The Line Plot with Multiple

Figure 6–21
The Vector Graph

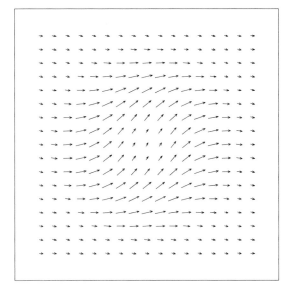

Straight Line appears plotting columns 5 through 10 as XY pairs.

10. To plot the vector data using SigmaPlot, create a Line Plot with Multiple Straight Line graph that plots columns 5 through 10 as three vector XY column pairs.

For more information on creating graphs in SigmaPlot, see the ***SigmaPlot User's Manual***.

Vector Transform
(VECTOR.XFM)

```
'   ******** VECTOR PLOT TRANSFORM ********
' Given a field of vector x,y positions, angles
' and lengths (in four columns), this transform
' will generate six columns of data that can be
' plotted to display the original data as vectors
' with arrow heads.
' The input data is located in columns xc to xc+3
' with x,y in columns xc and xc+1, vector angles
' in column xc+2 and vector lengths in column xc+3.
' The results are placed in columns xc+4 to xc+9
' To generate the vector plot, make a Line Plot
' with Multiple Straight Lines using XYpairs of
' these columns: xc+4 vs xc+5, xc+6 vs xc+7,
' xc+8 vs xc+9.
' This transform may be used in conjunction with the
' MESH.XFM transform which generates x,y pairs and
' corresponding z values.
pi= 3.14159265359
```

```
'        *********** Input ************
xc=1          ' column for start of data block
L =.1         ' length of arrow head
Angle = pi/6    ' angle of arrow head (radians).
' ********** Constant Vector Length **********
' To specify a constant vector length uncomment the
' two lines below and specify the vector length.
l=.5  ' length of vector (used only for constant
      ' length vectors). Uncomment the two
      ' statements below to use this value to specify
      ' vectors with constant length l.
      ' This will overwrite any data in column xc+3.
'nm=size(col(xc))
'col(xc+3)=data(1,1,nm)
'     *********** Results ************
' Column numbers for the vector output. These
' two columns will contain the data that displays
' the body of each vector.
body_x = xc+4
body_y = xc+5
' Columns containing the coordinates of the
' "left-hand" branch of the arrow head.
left_branch_x = xc+6
left_branch_y = xc+7
' Columns containing the coordinates of the
' "right-hand" branch of the arrow head.
right_branch_x = xc+8
right_branch_y = xc+9
'     *********** Program ************
x=col(xc)           'x positions of the vector field
y=col(xc+1)         'y positions of the vector field
theta=col(xc+2)     'angles of the vectors
m=abs(col(xc+3))    'lengths of the vectors
start_x=x-(m/2)*cos(theta)
start_y=y-(m/2)*sin(theta)
end_x=x+(m/2)*cos(theta)
end_y=y+(m/2)*sin(theta)
' Calculate the coordinates of the bodies of
' the vectors.
sx = data(1,size(end_x)*3)
col(body_x) = if(mod(sx,3)=1,start_x[int(sx/3)+1],
      if(mod(sx,3)=2,end_x[int(sx/3)+1],0/0))
col(body_y) = if(mod(sx,3)=1,start_y[int(sx/3)+1],
      if(mod(sx,3)=2,end_y[int(sx/3)+1],0/0))
' Calculate the coordinates of the arrow heads for
' the vectors.
vec_col_size = size(col(body_x))
 for i = 1 to vec_col_size step 3 do
 temp = if(cell(body_x,i)=cell(body_x,i+1),pi/2,
```

```
    arctan((cell(body_y,i+1) - cell(body_y,i))/
      (cell(body_x,i+1)-cell(body_x,i))))
 Theta = if(cell(body_y,i)-cell(body_y,i+1)<0,
  if(cell(body_x,i) <= cell(body_x,i+1),temp,
 if(cell(body_x,i) > cell(body_x,i+1),pi+temp,0/0)),
 if(cell(body_x,i) < cell(body_x,i+1),temp,
 if(cell(body_x,i) >= cell(body_x,i+1),pi+temp,0/0)))
 cell(left_branch_x,i) = cell(body_x,i+1)
 cell(left_branch_x,i+1) = L * cos(pi + Theta
  - Angle) + cell(body_x,i+1)
 cell(left_branch_x,i+2) = 0/0
 cell(left_branch_y,i) = cell(body_y,i+1)
 cell(left_branch_y,i+1) = L * sin(pi + Theta
  - Angle) + cell(body_y,i+1)
 cell(left_branch_y,i+2) = 0/0
 cell(right_branch_x,i) = cell(body_x,i+1)
 cell(right_branch_x,i+1) = L * cos(pi + Theta
  + Angle) + cell(body_x,i+1)
 cell(right_branch_x,i+2) = 0/0
 cell(right_branch_y,i) = cell(body_y,i+1)
 cell(right_branch_y,i+1) = L * sin(pi + Theta
  + Angle) + cell(body_y,i+1)
 cell(right_branch_y,i+2) = 0/0
 end for
```

Z Plane Design Curves

The ZPLANE.XFM transform is a specific example of the use of transforms to generate data for a unit circle and curves of constant damping ratio and natural frequency.

The root locus technique analyzes performance of a digital controller in the z plane using the unit circle as the stability boundary and the curves of constant damping ratio and frequency for a second order system to evaluate controller performance.

Root locus data is loaded from an external source and plotted in Cartesian coordinates along with the design curves in order to determine performance.

Refer to *Digital Control of Dynamic Systems*, Gene. F. Franklin and J. David Powell, Addison-Wesley, pp. 32 and 104 for the equations and graph.

To calculate the data for the design curves, you can either use the provided sample data and graph, or begin a new notebook, enter your own data, and create your own graph using the data.

1. To use the sample worksheet and graph, double-click the graph page icon in the Z Plane section of the Transform Examples notebook. The Z Plane worksheet appears with data in columns 1 through 10. The Z Plane graph page appears with the design curve data plotted over some sample root locus data. This plot uses columns 1 and 2 as the first curve and columns 3 and 4 as the second curve.

2. To use your own data, place your root locus, zero, and pole data in columns 1 through 10. If your locus data has been placed in other columns, you can change the location of the results columns after you open the ZPLANE.XFM file.

3. To plot the design curves of your data, create a Line Plot with Multiple Spline Curves, then plot column 1 as the X data against column 2 as the Y data for the first curve and column 3 as the X data against column 4 as the Y data as the second curve.

4. Press F10 to open the User-Defined Transform dialog box, then click the Open button, and open the ZPLANE.XFM transform in the XFMS directory. If necessary, change the res variable to the correct column number.

5. Click Run. The results are placed in columns 11 through 20, or the columns specified by the res variable.

6. If you opened the Z Plane graph, view the page. The circle, frequency trajectory, and damping trajectory data is automatically plotted with the design data.

7. To plot the circle data using SigmaPlot, create Multiple Line Plots with Simple Spline Curves. For the first plot use column 11 as the X values versus column 12 as the Y values. To plot the frequency trajectory data (zeta) plot column 13 versus column 14 and column 15 versus column 16 as the XY pairs. To plot the damping trajectory data (omega) plot column 17 versus column 18 and column 19 versus column 20 as the XY pairs.

For more information on creating graphs in SigmaPlot, see the SigmaPlot *User's Manual*.

Figure 6–22
Z Plane Graph

Root Locus for Compensated Antenna Design

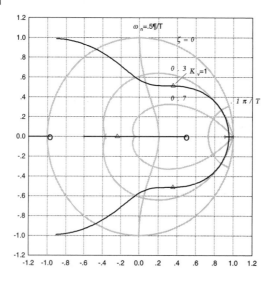

Z Plane Transform (ZPLANE.XFM)

```
'****** Transform for Z Plane Design Curves ******
' This transform generates the data for a unit
' circle and curves of constant damping ratio and
' natural frequency
' See Digital Control of Dynamic Systems,
' G.F. Franklin, J.D. Powell, pp. 32, 104.
' Root locus, zero, and pole data is loaded from an
' external source
res=11
'********** CALCULATE DATA FOR UNIT CIRCLE ********
pi=3.1415926
n=50
theta=data(0,2*pi*(1+1/n),pi/n)
circ_x=cos(theta)        'circle x coordinates
circ_y=sin(theta)        'circle y coordinates
'**** CALCULATE CONSTANT DAMPING TRAJECTORIES ****t
th=data(0,pi*(1+1/n),pi/n)
'Data for zeta 1:
z1=.3               'zeta 1 constant
r1=exp(-th*z1/(sqrt(1-z1^2)))
z1_x={r1*cos(th),"--",r1*cos(th)} 'zeta 1 x coord
z1_y={r1*sin(th),"--",-r1*sin(th)} 'zeta 1 y coord
'Data for zeta 2:
z2=.7               'zeta 2 constant
r2=exp(-th*z2/(sqrt(1-z2^2)))
z2_x={r2*cos(th),"--",r2*cos(th)} 'zeta 2 x coord
z2_y={r2*sin(th),"--",-r2*sin(th)} 'zeta 2 y coord
'*** CALCULATE CONSTANT FREQUENCY TRAJECTORIES ***
'Data for omega 1:
wnT1=0.1*pi          'omega 1 constant
z={data(0,.99,.05),.9999}
th1=wnT1*sqrt(1-z^2)
r3=exp(-th1*z/(sqrt(1-z^2)))
w1_x={r3*cos(th1),"--",r3*cos(th1)}   'omega 1 x
w1_y={r3*sin(th1),"--",-r3*sin(th1)}  'omega 1 y
'Data for omega 2:
wnT2=.5*pi             'omega 1 constant
th2=wnT2*sqrt(1-z^2)
r4=exp(-th2*z/(sqrt(1-z^2)))
w2_x={r4*cos(th2),"--",r4*cos(th2)}   'omega 2 x
w2_y={r4*sin(th2),"--",-r4*sin(th2)}  'omega 2 y
'* PLACE CIRCLE AND TRAJECTORY DATA IN WORKSHEET *
col(res)=circ_x      'circle x coordinates
col(res+1)=circ_y     'circle y coordinates
col(res+2)=z1_x      'zeta 1 x coordinates
col(res+3)=z1_y      'zeta 1 y coordinates
col(res+4)=z2_x      'zeta 2 x coordinates
col(res+5)=z2_y      'zeta 2 y coordinates
col(res+6)=w1_x      'omega 1 x coordinates
```

Introduction To The Regression Wizard

This chapter describes:

➤ An overview of regression.
➤ The Regression Wizard (see page 142).
➤ Opening .FIT files (see page 143).
➤ The Marquardt-Levenberg curve fitting algorithm (see page 145).

Regression Overview

What is Regression?   *Regression* is most often used by scientists and engineers to visualize and plot the curve that best describes the shape and behavior of their data.

Regression procedures find an association between independent and dependent variables that, when graphed on a Cartesian coordinate system, produces a straight line, plane or curve. This is also commonly known as curve fitting.

The *independent* variables are the known, or predictor, variables. These are most often your X axis values. When the independent variables are varied, they result in corresponding values for the *dependent*, or response, variables, most often assigned to the Y axis.

Regression finds the equation that most closely describes, or fits, the actual data, using the values of one or more independent variables to predict the value of a dependent variable. The resulting equation can then be plotted over the original data to produce a curve that fits the data.

The Regression Wizard

SigmaPlot uses the Regression Wizard to perform regression and curve fitting. The Regression Wizard provides a step-by step guide through the procedures that let you fit the curve of a known function to your data, and then automatically plot the curve and produce statistical results.

The Regression Wizard greatly simplifies curve fitting. There is no need to be familiar with programming or higher mathematics. The large library of built-in equations are graphically presented and organized by different categories, making selection of your models very straight-forward. Built-in shortcuts let you bypass all but the simplest procedures; fitting a curve to your data can be as simple as picking the equation to use, then clicking a button.

The Regression Wizard can be used to

➤ Select the function describing the shape of your data. SigmaPlot provides over 100 built-in equations. You can also create your own custom regression equations.

➤ Select the variables to fit to the function. You can select your variables from either a graph or a worksheet.

➤ Evaluate and save your results. Resulting curves can be plotted automatically on a graph, and statistical results saved to the worksheet and text reports.

These procedures are described in further detail in the next chapter.

Figure 7–1
Selecting an Equation from
the Regression Wizard

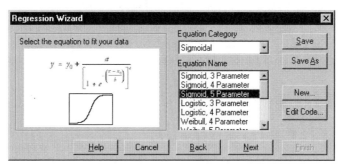

Feature Highlights	The Regression Wizard offers notable improvements over the previous SigmaPlot curve fitters. These include:

➤ A built-in Equation Library that can be extended limitlessly with your own user-defined functions and libraries.

➤ Graphical display of built-in equations.

➤ Graphical selection of variables from worksheets or graphs.

➤ Automatic parameter estimation for widely varied datasets.

➤ Automatic plotting of results.

➤ Automatic generation of textual reports.

➤ Inclusion of fit equations into notebooks.

The Regression Wizard is also one-hundred percent compatible with older .FIT files, as described below.

Opening .FIT Files

Use the File menu Open command to open old curve fit (.FIT) files, selecting SigmaPlot Curve Fit as the file type. .FIT files are opened as a single equation in a notebook.

.FIT files can also be opened from the library panel of the Regression Wizard.

Adding .FIT Files to a Library or Notebook	You can add these equations to other notebooks by copying and pasting. To add them to your regression library, open the library notebook (STANDARD.JFL for SigmaPlot's built-in library), then copy the equation and paste it into the desired section of the library notebook.

You can also create your own library by simply combining all your old .FIT files into a single notebook, then setting this notebook to be your default equation library (see "Using a Different Library for the Regression Wizard" on page 176).

Remember, sections appear as categories in the library, so create a new section to create a new equation category.

Figure 7–2
Opening a .FIT file as a
notebook using the File
menu Open... command

.FIT files as well as new equations do not have graphic previews of the equation.

About the Curve Fitter

The curve fitter works by varying the parameters (coefficients) of an equation, and finds the parameters which cause the equation to most closely fit your data. Both the equation and initial parameter values must be provided. All built-in equations have the curve equation and initial parameters predefined.

The curve fitter accepts up to 25 equation parameters and ten independent equation variables. You can also specify up to 25 parameter constraints, which limit the search area of the curve fitter when checking for parameter values.

The regression curve fitter can also use weighted least squares for greater accuracy.

Curve-fitting Algorithm

The SigmaPlot curve fitter uses the Marquardt-Levenberg algorithm to find the coefficients (parameters) of the independent variable(s) that give the "best fit" between the equation and the data.

This algorithm seeks the values of the parameters that minimize the sum of the squared differences between the values of the observed and predicted values of the dependent variable where y_i is the observed and $\hat{y}_i$ is the predicted value of the

$$SS = \sum_{i=1}^{n} w_i(y_i - \hat{y}_i)^2$$

dependent variable.

This process is *iterative*—the curve fitter begins with a "guess" at the parameters, checks to see how well the equation fits, then continues to make better guesses until the differences between the residual sum of squares no longer decreases significantly. This condition is known as *convergence*.

For informative references about curve-fitting algorithms, see below.

References for the Marquardt-Levenberg Algorithm

Press, W. H., Flannery, B. P., Teukolsky, S. A., and Vetterling, W. T. (1986). *Numerical Recipes*. Cambridge: Cambridge University Press.

Marquardt, D.W. (1963). An Algorithm for Least Squares Estimation of Parameters. *Journal of the Society of Industrial and Applied Mathematics*, 11, 431-441.

Nash, J.C. (1979). *Compact Numerical Methods for Computers: Linear Algebra and Function Minimization*. New York: John Wiley & Sons, Inc.

Shrager, R.I. (1970). Regression with Linear Constraints: An Extension of the Magnified Diagonal Method. *Journal of the Association for Computing Machinery*, 17, 446-452.

Shrager, R.I. (1972). Quadratic Programming for N. *Communications of the ACM*, 15, 41-45.

8 Regression Wizard

The Regression Wizard is designed to help you select an equation and other components necessary to run a regression on your data. The Regression Wizard guides you through:

➤ Selecting your equation, variables, and other options.

➤ Saving the results and generating a report.

➤ Plotting the predicted variables.

Using the Regression Wizard

To run the Regression Wizard:

Selecting the Data Source

1. Open or view the page or worksheet with the data you want to fit.

   If you select a graph, right-click the curve you want fitted, and choose Fit Curve.

   If you are using a worksheet, highlight the variables you want to fit, then press F5 or choose the Statistics menu Regression Wizard command. The Regression Wizard opens.

Selecting the Equation to Use

2. Select an equation using the Equation Category and Equation Name drop-down lists. You can view different equations by selecting different categories and names. The equation's mathematical expression and shape appear to the left.

Figure 8–1
Selecting an Equation
Category and
Equation Name

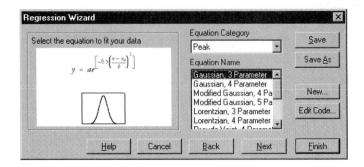

For a complete list of the built-in equations, see "Regression Equation Library" on page 283.

If the equation you want to use isn't on this list, you can create a new equation. See "Editing Code" on page 183 for more information. You can also browse other notebooks and regression equation libraries for other equations; see "Regression Equation Libraries and Notebooks" on page 175 for more information on using equation libraries.

Σ Note that the equation you select is remembered the next time you open the wizard.

If the Finish button is available, you can click it to complete your regression. If it is not available, or if you want to further specify your results, click Next.

Selecting the Variables to Fit

3. Clicking Next opens the variables panel. You can select or reselect your variables from this panel. To select a variable, click a curve on a graph, or click a column in a worksheet. The equation picture to the left prompts you for which variable to select.

You can also modify other equation settings and options from this panel using the Options button. These options include changing initial parameter estimates, parameter constraints, weighting, and other related settings.

For more information on setting options, see "Equation Options" on page 155.

Figure 8–2
Selecting a Plot as the
Data Source for the
Regression Wizard

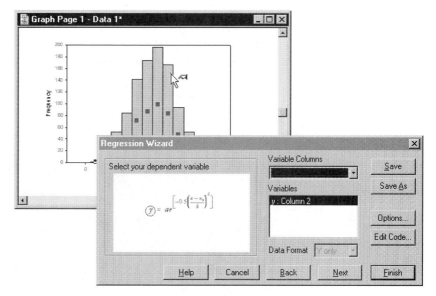

If you pick variables from a worksheet column, you can also set the data format. See "Variable Options" below for descriptions of the different data formats.

Viewing Initial Results

4. When you have selected your variables, you can either click Finish, or click Next to continue.

   Clicking Next executes the regression equation and displays the initial results. These results are also displayed if you receive a warning or error message about your fit.

Figure 8–3
The Initial Results for
a Regression

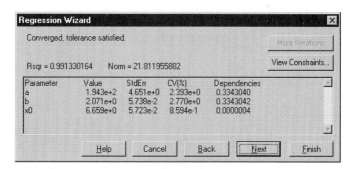

For interpretation of these results, see "Interpreting Initial Results" on page 163.

Setting Results Options

5. If you wish to modify the remainder of the results that are automatically saved, click Next. Otherwise, click Finish.

The first results panel lists

➤ which results are saved to the worksheet

➤ whether or not a text report of the regression is to be generated

➤ whether or not a copy of the regression equation is saved to the section that contains the data that was fitted

6. Select which results you want to keep. These settings are remembered between regression sessions.

Figure 8–4
Selecting the
Results to Save

These settings are retained
between sessions.

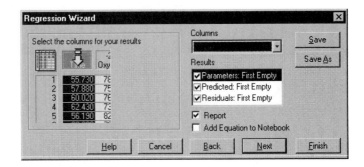

7. Click Finish or click Next to select the graphed results.

Setting Graph Options

8. If you selected your variables from a graph, you can add your equation curve to that graph automatically. You can also plot the equation on any other graph on that page.

You also always have the option of creating a new graph of the original data and fitted curve.

Figure 8–5
Selecting the
Results to Graph

These settings are retained
between sessions.

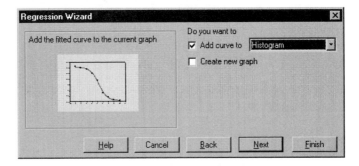

After selecting the graphed results you want, click Finish. Click Next only if you want to select the specific columns used to contain the data for the equation.

9. To select the columns to use for the plotted results, click the columns in the worksheet where you want the results to always appear. Remember, these settings are re-used each time you perform a regression.

Figure 8–6
Selecting the Graph
Results Columns

These settings are retained
between sessions.

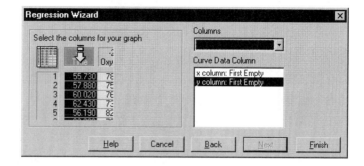

Finishing the Regression

When you click Finish, all your results are displayed in the worksheet, report, and graph. The initial defaults are to save parameter and computed dependent variable values to the worksheet, to create a statistical report, and to graph the results.

To change the results that are saved, use the Next button to go through the entire wizard, changing your settings as desired.

Running a Regression From a Notebook

Because regression equations can be treated like any other notebook item, you can select and open regression equations directly from a notebook. This is particularly convenient if you have created or stored equations along with the rest of your graphs and data.

1. **View the note**book with the equation you want to use, and double-click the equation. You can also click the equation, then click the Open button. The Regression Wizard opens with the equation selected.

2. Select the variables as prompted by clicking a curve or worksheet columns. Note that at this point you can open and view any notebook, worksheet or page you would like, and pick your variables from that source.

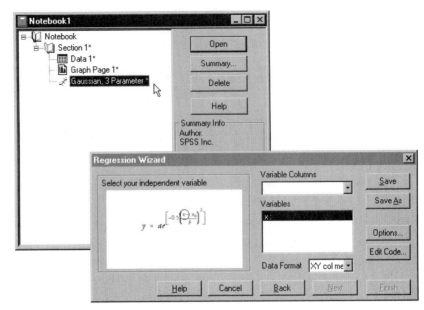

Figure 8–7
Selecting a Regression
Equation from a Notebook
to Start the Regression
Wizard

3. Click Finish to complete the regression, or click Next if you want to view initial results or change your results options.

Creating New Regression Equations

You can create new regression equations two different ways: by using the New button in the Regression Wizard, or by creating a new item for a notebook.

When you create a new equation, the Regression editing dialog box appears with blank headings. For information on how to fill in these headings, see "Editing Code" on page 183.

Viewing and Editing Code

Viewing Code

To view the code for the current equation document, click the Edit Code button. For more information, see "Editing Code" on page 183.

You can click the Edit Code button from the equation or variables panels. The Edit Code button opens the Regression dialog box. All settings for the equation are displayed.

Figure 8–8
Viewing the code for a
built-in equation in the
Regression dialog box.

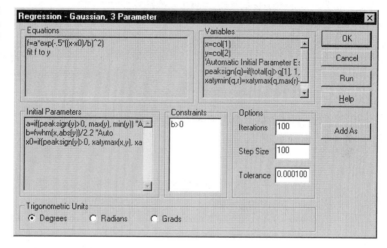

Σ
Note that the Equations, Parameters, and Variables are non-editable for built-in SigmaPlot equations. However, you can edit and save our built-in equations as new equations. Simply click Add As, add the equation to the desired section, and then edit the Equations, Variables and Parameters as desired.

You can also copy and paste equations from notebook to notebook like any other notebook item. Pasted built-in equations also become completely editable.

Entering the code for new equations is described in detail in "Editing Code" on page 183.

Variable Options

Data Format Options

If you use data columns from the worksheet, you can specify the data format to use in the variables panel. By default, the data format when assigning columns from the worksheet is XY Pair.

Figure 8–9
Variable Data
Format Options

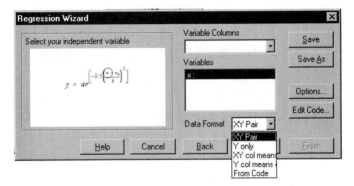

The data format options are:

➤ XY pair: Select an x and a y variable.

➤ Y only: Select only a y variable column.

➤ XY column means: pick one x column, then multiple y columns; the y columns will be graphed as means.

➤ Y column means only: pick multiple y columns; the columns will be graphed as means.

➤ From Code: uses the current settings as shown when editing code.

When you use an existing graph as your data source, the Regression Wizard displays a format reflecting the data format of the graph. You cannot change this format unless you switch to using the worksheet as your data source, or run the regression directly from editing the code.

Multiple Independent
Variables

Although the standard regression library only supports up to two independent variables, the curve fitter can accept up to ten. To use models that have more than two independent variables, simply create or open a model with the desired equation and variables. The Regression Wizard will prompt you to select columns for each defined variable.

Figure 8–10
Variable Data Format Options
for a 3D Function

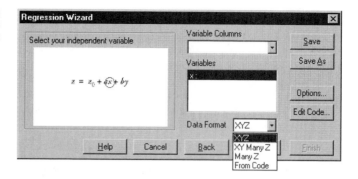

Equation Options

If the curve fitter fails to find a good fit for the curve, you can try changing the regression options to see if you can improve the fit. To set options for a regression, click the Options button in the Variables panel of the Regression Wizard. The Regression Options dialog box of the Regression Wizard appears.

Figure 8–11
Regression Option
Dialog Box

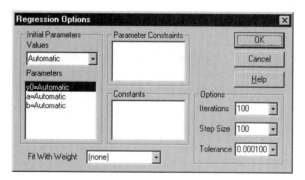

Σ If you want to edit the settings in the equation document manually, click the Edit Code button. For more information on editing equation documents manually, see "Editing Code" on page 183.

Use the Regression Options dialog box to

➤ change initial parameter values

➤ add or change constraints

➤ change constant values

> ➤ use weighted fitting, if it is available
> ➤ change convergence options

Parameters   The default setting for the equation is shown. The Automatic setting available with the built-in SigmaPlot equations uses algorithms that analyze your data to predict initial parameter estimates. These do not work in all cases, so you may need to enter a different value. Just click the parameter you want to change, and make the change in the edit box.

The values that appear in the Initial Parameters drop-down list were previously entered as parameter values. Any parameter values you enter will also be retained between sessions.

Figure 8–12
Setting Initial
Parameter Options

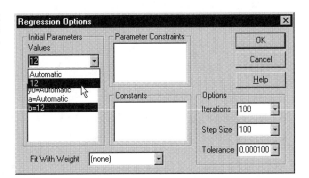

Parameters can be either a numeric value or a function. The value of the parameter should approximate the final result, in order to help the curve fitter reach a valid result, but this depends on the complexity and number of parameters of the equation. Often an initial parameter nowhere near the final result will still work. However, a good initial estimate helps guarantee better and faster results.

For more information on how parameters work, see "Initial Parameters" on page 196. For an example on the effect of different initial parameter values, see "Curve Fitting Pitfalls" on page 217. For more information on the use of automatic parameter estimation, see "Automatic Determination of Initial Parameters" on page 200.

Constraints   Constraints are used to set limits and conditions for parameter values, restricting the regression search range and improving curve fitter speed and accuracy. Constraints are often unnecessary, but should always be used whenever appropriate for your model.

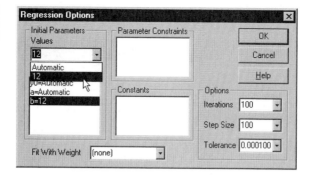

Figure 8–13
Setting Initial
Parameter Options

Constraints are also useful to prevent the curve fitter from testing unrealistic parameter values. For example, if you know that a parameter should always be negative, you can enter a constraint defining the parameter to be always less than 0.

You can also use constraints if the regression produces parameter values that you know are inaccurate. Simply click Back from the initial results panel, click the Options button, and enter constraint(s) that prevent the wrong parameter results.

Note that if the curve fitter encounters constraints while iterating, you can view these constraints from the initial results panel using the Constraints button. For more information, see "Checking Use of Constraints" on page 164.

Entering Parameter Constraints

To enter constraints, click the Constraints edit box, and type the desired constraint(s), using the transform language operators.

A constraint must be a linear equation of the equation parameters, using an equal ($=$) or inequality ($<$ or $>$) sign. For example, you could enter the following constraints for the parameters a, b, c, d, and e:

$a<1$
$10_*b+c/20>2$
$d-e=15$
$a>b+c+d+e$

However, the constraint

$a_*x<1$

is illegal, since x is a variable, not a parameter, and the constraints

$b+c^2>4$
$d_*e=1$

are illegal because they are nonlinear. Inconsistent and conflicting constraints are automatically rejected by the curve fitter.

Figure 8–14
Entering Parameter
Constraints

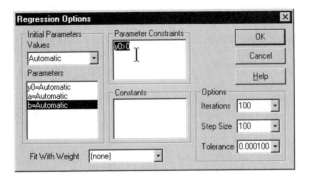

Defining Constants

Constants that appear in the Constants edit window have been previously defined as a constant, rather than a parameter to be determined by the regression. To edit a constant value, or define new constant values, use the Edit Code... option of the Wizard dialog box. For more information on editing and defining new constant values, see "Defining Constants" on page 188.

Constants are defined when an equation is created. Currently, you can only define new constants by editing the regression equation code.

However, you can redefine any existing constants. Change only the value of the constant. Do not add new constant values; constant variables must exist in the equation and not be defined already under variables or parameters, so they can only be defined within the code of an equation.

Fit with Weight

You can select from any of the weights listed. Some built-in equations have some predefined values, although most do not. If no weighting options are available for your equation, only the None option will be available.

Weighting options appear in the Fit with Weight drop-down list. By default, the weighting applied to the fit is None. To apply a different weighting setting, select a weighting option from the drop-down list.

Figure 8–15
Selecting a Predefined
Weight Variable

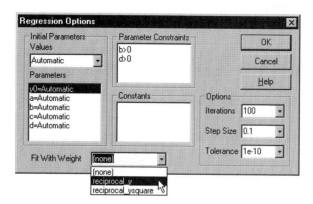

Weight variables must be defined by editing the regression code. For information on how to define your own weighting options, see "Weight Variables" on page 195.

For a demonstration of weighting variable use, see "Example 2: Weighted Regression" on page 223.

Iterations

The Iterations option sets the maximum number of repeated fit attempts before failure. Each iteration of the curve fitter is an attempt to find the parameters that best fit the model. With each iteration, the curve fitter varies the parameter values incrementally, and tests the fit of that model to your data. When the improvement in the fit from one iteration to the next is smaller than the setting determined by the Tolerance option, the curve fitter stops and displays the results.

Figure 8–16
Changing Iterations

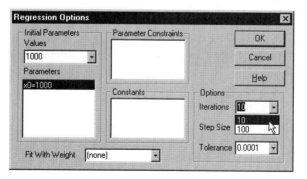

Changing the number of iterations can be used to speed up or improve the regression process, especially if more than the default of 100 iterations are required for a complex fit. You can also reduce the number of iterations if you want to end a fit to check on its interim progress before it takes too many iterations.

To change the maximum number of iterations, enter the number of iterations to use, or select a previously used number of iterations from the drop down list.

When the maximum number of iterations is reached, the regression stops and the current results are displayed in the initial parameters panel. If you want to continue with more iterations, you can click the Iterations button. For more information on using the Iterations button, see "More Iterations" on page 164.

For more information on the use of iterations, see "Iterations" on page 199.

Step Size Step size, or the limit of the initial change in parameter values used by the curve fitter as it tries, or *iterates,* different parameter values, is a setting that can be changed to speed up or improve the regression process.

Figure 8–17
Changing Step Size

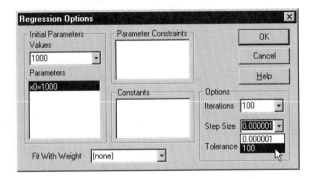

A large step size can cause the curve fitter to wander too far away from the best parameter values, whereas a step size that is too small may never allow the curve fitter to reach the value of the best parameters.

The default step size value is 100. To change the Step Size value, type the desired step size in the Step Size edit box, or select a previously defined value from the drop-down list.

For more information on use of the Step Size option, see "Step Size" on page 199.

For an example of the possible effects of different step sizes, see "Curve Fitting Pitfalls" on page 217.

Tolerance The Tolerance option controls the condition that must be met in order to end the regression process. When the absolute value of the difference between the *norm* of the residuals (square root of the sum of squares of the residuals), from one iteration to the next, is less than the tolerance value, the iteration stops. The norm for each iteration is displayed in the progress dialog box, and the final norm is displayed in the initial results panel.

Figure 8–18
Changing Tolerance

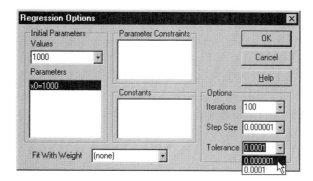

When the tolerance condition has been met, a minimum of the sum of squares has usually been found, which indicates a correct solution. However, local minimums in the sum of squares can also cause the curve fitter to find an incorrect solution. For an example of the possible effects of different tolerance values, see "Curve Fitting Pitfalls" on page 217.

Decreasing the value of the tolerance makes the requirement for finding an acceptable solution more strict; increasing the tolerance relaxes this requirement.

The default tolerance setting is 0.0001. To change the tolerance value, type the desired value in the Tolerance edit box, or select a previously defined value from the drop-down list. For more details on the use of changing tolerance, see "Tolerance" on page 199.

Saving Regression Equation Changes

When an equation is edited using the Options or Regression dialogs, or when you add an equation, all changes are updated to the equation in the library or notebook. However, just like other notebook items, these changes are not saved to the file until the notebook is saved. Changes made to regression libraries are automatically saved when the Regression Wizard is closed.

You can also save changes to regression libraries using the Save or Save As buttons in the Regression Wizard. This saves the current regression library notebook to disk. Save As allows you to save the regression library to a new file.

If you have a regression library open as a notebook, you can also save changes by saving the notebook using the File menu Save or Save As... command.

Watching The Fit Progress

During the regression process, the Regression fit progress dialog box displays the number of iterations completed, the norm value for each iteration, and a progress bar indicating the percent complete of the maximum iterations.

Figure 8–19
The Regression Fit
Progress Dialog Box

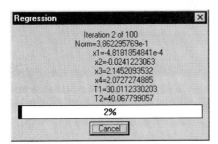

Cancelling a Regression

To stop a regression while it is running, click the Cancel button. The initial results appear, displaying the most recent parameter values, and the norm value. You can continue the regression process by clicking the More Iterations button.

Interpreting Initial Results

When you click Next from the variables panel, the regression process completes by either converging, reaching the maximum number of iterations, or encountering an error. When any of these conditions are met, or whenever there is an error or warning, the initial results panel is displayed.

Figure 8–20
Initial Regression Results

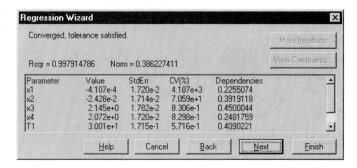

Completion Status Messages A message displaying the condition under which the regression completed is displayed in the upper left corner of the Regression Wizard. If the regression completed with convergence, the message:

`Converged, tolerance satisfied`

is displayed. Otherwise, another status or error message is displayed. For a description of these messages, see "Regression Results Messages" on page 180.

Rsqr R^2 is the ***coefficient of determination***, the most common measure of how well a regression model describes the data. The closer R^2 is to one, the better the independent variables predict the dependent variable.

R^2 equals 0 when the values of the independent variable does not allow any prediction of the dependent variables, and equals 1 when you can perfectly predict the dependent variables from the independent variables.

Initial Results The initial results are displayed in the results window, in five columns.

Parameter The parameter names are shown in the first column. These parameters are derived from the original equation.

Value The calculated parameter values are shown in the second column.

StdErr The asymptotic standard errors of the parameters are displayed in column three. The standard errors and coefficients of variation (see next) can be used as a gauge of the fitted curve's accuracy.

CV(%) The parameter coefficients of variation, expressed as a percentage, are displayed in column four. This is the normalized version of the standard errors:

$$CV\% = standard\ error \times 100 \ \S\ parameter\ value$$

The coefficient of variation values and standard errors (see above) can be used as a gauge of the accuracy of the fitted curve.

Dependency The last column shows the parameter dependencies. The dependence of a parameter is defined to be

$$dependence = 1 - \frac{(variance\ of\ the\ parameter,\ other\ parameters\ constant)}{(variance\ of\ the\ parameter,\ other\ parameters\ changing)}$$

Parameters with dependencies near 1 are strongly dependent on one another. This may indicate that the equation(s) used are too complicated and "over-parameterized"—too many parameters are being used, and using a model with fewer parameters may be better.

Changing the Regression Equation or Variables

To go back to any of the previous panels, just click Back. This is especially useful if you need to change the model (equation) used, or if you need to modify any of the equation options and try the curve fit again.

More Iterations

If the maximum number of iterations was reached before convergence, or if you canceled the regression, the More Iterations button is available. Click More Iterations to continue for as many iterations as specified by the Iterations equation option, or until completion of the regression.

Checking Use of Constraints

If you used parameter constraints, you can determine if the regression results involved any constraints by clicking the View Constraints button. This button is dimmed if no constraints were entered.

Figure 8–21
The Constraints Dialog Box

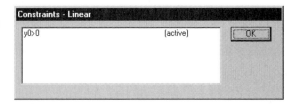

The Constraints dialog box displays all constraints, and flags the ones encountered with the word "(active)". A constraint is flagged as active when the parameter values lie on the constraint boundary. For example, the constraint:

a+b<1

is active when the parameters satisfy the condition a+b=1, but if a+b<1, the constraint is inactive.

Note that an equality constraint is always active (unless there are constraint inconsistencies).

Quitting the Regression
If the regression results are unsatisfactory, you can click Back and change the equation or other options, or you can select Cancel to close the wizard.

If you want to keep your results, click Finish. You can also click Next to specify which results you want to keep.

Saving Regression Results

Regression reports and other data results are saved using the Regression Wizard results options panel, which appears after the initial results panel. Settings made here are retained from session to session.

The type of data results that can be saved to the current notebook for each regression procedure are

➤ the function results, saved to the worksheet

➤ a statistical report

➤ a copy of the regression equation

Saving the Results using Default Settings
To save the regression results using the default save setting, click Finish at any point the Finish button is active. If you want to see or modify the results that are produced, you can use the Next button to advance to the results options panel.

Saving Results to the Worksheet
Function results can be saved to the current worksheet. These are

➤ equation parameter values

➤ predicted values of the dependent variable for each independent variable value

➤ residuals, or the difference between the predicted and observed dependent variable values

To place any of these values in a column in the worksheet, simply check the results you want to keep. If you want to set a specific column in which to always place these values, you can click a column on a worksheet for each result.

Figure 8–22
Generating and Saving
a Report from the
Regression Wizard

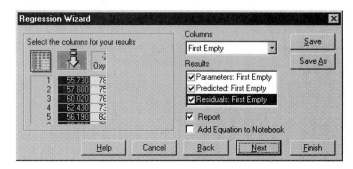

Saving a Report

Regression reports are saved to the current section by checking the Report option. For more information about interpreting reports, see "Interpreting Regression Reports" on page 167.

Adding the Equation
to the Notebook

To add the current regression equation to the current notebook, check the Add Equation to Notebook check box.

If this option is selected, a copy of the equation is added to the current section of your notebook.

Graphing Regression Equations

SigmaPlot can graph the results of a regression as a fitted curve. A curve or graph is created by default. If you want to disable graphed results, you can change the options in the Regression Wizard graph panel. Note that these settings are retained from session to session.

From the graph panel, you can choose to plot the results either by

➤ adding a plot to an existing graph. This option is only available if the fitted variables were assigned by selecting them from a graph.

➤ creating a new graph of the original data and fitted curve

To add a plot to an existing graph, click the Add Curve to check box option, then select the graph to which you want to add a plot from the drop-down list. The drop-down list includes all the graphs on the current page. If there is no existing graph, this option is dimmed.

If you want to specify the columns used to plot the fitted curve, click Next. Otherwise, the data is placed in the first available columns.

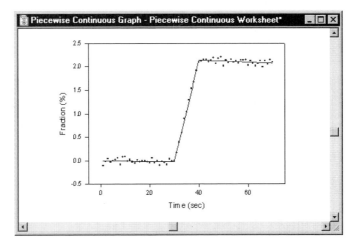

Figure 8–23
A Fitted Curve
Added to a Graph

To create a new graph, click the Create New Graph check box. Click Finish to create a new notebook section containing a worksheet of the plotted data and graph page.

Data Plotted for Regression Curves

You can specify the worksheet columns used to add a fitted curve to an existing graph, or to create a new graph, by clicking Next from the graph panel.

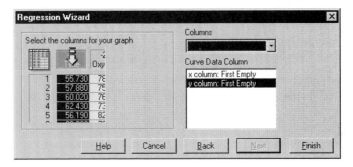

Figure 8–24
The Regression Wizard
Pick Output Dialog Box

From this panel you can select worksheet columns for X, Y, (and Z data for 3D graphs) by clicking worksheet columns. The default of First Empty places the results in the first available column after the last filled cell.

Interpreting Regression Reports

Reports can be automatically generated by the Regression Wizard for each curve fitting session. The statistical results are displayed to four decimal places of precision by default.

Reports are displayed using the SigmaPlot report editor. For information on modifying reports, see "Creating Reports" on page 342 in the *User's Manual*.

Equation Code	This is a printout of the code used to generate the regression results.

See "Editing Code" on page 183, for more information on how to read the code for a regression equation. |

Figure 8–25
Regression Report

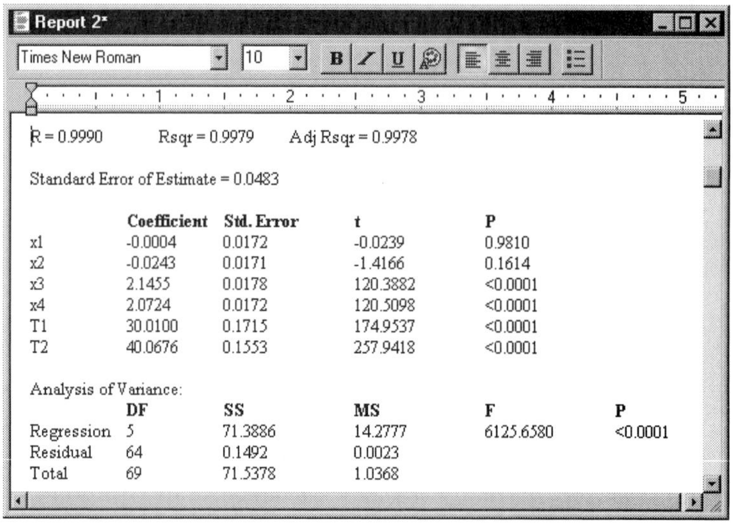

Report 2*

Times New Roman 10 **B** *I* U

R = 0.9990   Rsqr = 0.9979   Adj Rsqr = 0.9978

Standard Error of Estimate = 0.0483

	Coefficient	Std. Error	t	P
x1	-0.0004	0.0172	-0.0239	0.9810
x2	-0.0243	0.0171	-1.4166	0.1614
x3	2.1455	0.0178	120.3882	<0.0001
x4	2.0724	0.0172	120.5098	<0.0001
T1	30.0100	0.1715	174.9537	<0.0001
T2	40.0676	0.1553	257.9418	<0.0001

Analysis of Variance:

	DF	SS	MS	F	P
Regression	5	71.3886	14.2777	6125.6580	<0.0001
Residual	64	0.1492	0.0023		
Total	69	71.5378	1.0368		

R and R Squared	The multiple *correlation coefficient*, and R^2, the *coefficient of determination*, are both measures of how well the regression model describes the data. R values near 1 indicate that the equation is a good description of the relation between the independent and dependent variables.

R equals 0 when the values of the independent variable does not allow any prediction of the dependent variables, and equals 1 when you can perfectly predict the dependent variables from the independent variables. |
| **Adjusted R Squared** | The adjusted R^2, R^2_{adj}, is also a measure of how well the regression model describes the data, but takes into account the number of independent variables, which reflects the degrees of freedom. Larger R^2_{adj} values (nearer to 1) indicate that the equation is a good description of the relation between the independent and dependent variables. |
| **Standard Error of the Estimate ($S_{y|x}$)** | The standard error of the estimate $S_{y|x}$ is a measure of the actual variability about the regression plane of the underlying population. The underlying population generally falls within about two standard errors of the observed sample. |
| **Statistical Summary Table** | The standard error, t and P values are approximations based on the final iteration of the regression.

Estimate The value for the constant and coefficients of the independent variables for the regression model are listed. |

Standard Error The standard errors are estimates of the uncertainties in the estimates of the regression coefficients (analogous to the standard error of the mean). The true regression coefficients of the underlying population are generally within about two standard errors of the observed sample coefficients. Large standard errors may indicate multicollinearity.

t **statistic** The *t* statistic tests the null hypothesis that the coefficient of the independent variable is zero, that is, the independent variable does not contribute to predicting the dependent variable. *t* is the ratio of the regression coefficient to its standard error, or

$$t = \frac{regression\ coefficient}{standard\ error\ of\ regression\ coefficient}$$

You can conclude from "large" *t* values that the independent variable can be used to predict the dependent variable (i.e., that the coefficient is not zero).

P **value** *P* is the *P* value calculated for *t*. The *P* value is the probability of being wrong in concluding that the coefficient is not zero (i.e., the probability of falsely rejecting the null hypothesis, or committing a Type I error, based on *t*). The smaller the *P* value, the greater the probability that the coefficient is not zero.

Traditionally, you can conclude that the independent variable can be used to predict the dependent variable when $P < 0.05$.

Analysis of Variance (ANOVA) Table The ANOVA (analysis of variance) table lists the ANOVA statistics for the regression and the corresponding *F* value for each step.

SS (Sum of Squares) The sum of squares are measures of variability of the dependent variable.

➤ The sum of squares due to regression measures the difference of the regression plane from the mean of the dependent variable

➤ The residual sum of squares is a measure of the size of the residuals, which are the differences between the observed values of the dependent variable and the values predicted by the regression model

DF (Degrees of Freedom) Degrees of freedom represent the number of observations and variables in the regression equation.

➤ The regression degrees of freedom is a measure of the number of independent variables

➤ The residual degrees of freedom is a measure of the number of observations less the number of parameters in the equation

MS (Mean Square) The mean square provides two estimates of the population variances. Comparing these variance estimates is the basis of analysis of variance.

The mean square regression is a measure of the variation of the regression from the mean of the dependent variable, or

$$\frac{\textit{sum of squares due to regression}}{\textit{regression degrees of freedom}} \;=\; \frac{SS_{reg}}{DF_{reg}} \;=\; MS_{reg}$$

The residual mean square is a measure of the variation of the residuals about the regression plane, or

$$\frac{\textit{residual sum of squares}}{\textit{residual degrees of freedom}} \;=\; \frac{SS_{res}}{DF_{res}} \;=\; MS_{res}$$

The residual mean square is also equal to $S_{y|x}^{2}$.

F statistic The F test statistic gauges the contribution of the independent variables in predicting the dependent variable. It is the ratio

$$\frac{\textit{regression variation from the dependent variable mean}}{\textit{residual variation about the regression}} \;=\; \frac{MS_{reg}}{MS_{res}} \;=\; F$$

If F is a large number, you can conclude that the independent variables contribute to the prediction of the dependent variable (i.e., at least one of the coefficients is different from zero, and the "unexplained variability" is smaller than what is expected from random sampling variability of the dependent variable about its mean). If the F ratio is around 1, you can conclude that there is no association between the variables (i.e., the data is consistent with the null hypothesis that all the samples are just randomly distributed).

P value The P value is the probability of being wrong in concluding that there is an association between the dependent and independent variables (i.e., the probability of falsely rejecting the null hypothesis, or committing a Type I error, based on F). The smaller the P value, the greater the probability that there is an association.

Traditionally, you can conclude that the independent variable can be used to predict the dependent variable when $P < 0.05$.

PRESS Statistic

PRESS, the *Predicted Residual Error Sum of Squares*, is a gauge of how well a regression model predicts new data. The smaller the PRESS statistic, the better the predictive ability of the model.

Figure 8–26
Regression Report

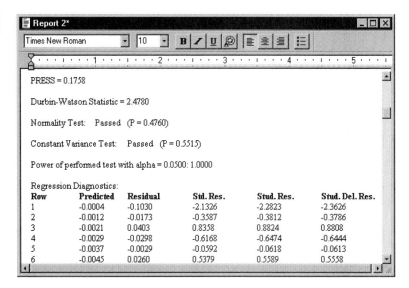

The PRESS statistic is computed by summing the squares of the prediction errors (the differences between predicted and observed values) for each observation, with that point deleted from the computation of the regression equation.

Durbin-Watson Statistic

The Durbin-Watson statistic is a measure of correlation between the residuals. If the residuals are not correlated, the Durbin-Watson statistic will be 2; the more this value differs from 2, the greater the likelihood that the residuals are correlated.

Regression assumes that the residuals are independent of each other; the Durbin-Watson test is used to check this assumption. If the Durbin-Watson value deviates from 2 by more than 0.50, a warning appears in the report, i.e., if the Durbin-Watson statistic is below 1.50 or above 2.50.

Normality Test

The normality test results display whether the data passed or failed the test of the assumption that the source population is normally distributed around the regression, and the *P* value calculated by the test. All regressions assume a source population to be normally distributed about the regression line. If the normality test fails, a warning appears in the report.

Failure of the normality test can indicate the presence of outlying influential points or an incorrect regression model.

Constant Variance Test

The constant variance test results displays whether or not the data passed or failed the test of the assumption that the variance of the dependent variable in the source population is constant regardless of the value of the independent variable, and the *P* value calculated by the test. When the constant variance test fails, a warning appears in the report.

If the constant variance test fails, you should consider trying a different model (i.e., one that more closely follows the pattern of the data) using a weighted regression, or transforming the independent variable to stabilize the variance and obtain more accurate estimates of the parameters in the regression equation.

If you perform a weighted regression, the normality and equal variance tests use the weighted residuals $w_i(y_i - \hat{y}_i)$ instead of the raw residuals $y_i - \hat{y}_i$.

Power

The power, or sensitivity, of a regression is the probability that the model correctly describes the relationship of the variables, if there is a relationship.

Regression power is affected by the number of observations, the chance of erroneously reporting a difference α (alpha), and the slope of the regression.

Alpha (a) Alpha (α) is the acceptable probability of incorrectly concluding that the model is correct. An α error is also called a *Type I error* (a Type I error is when you reject the hypothesis of no association when this hypothesis is true).

Smaller values of α result in stricter requirements before concluding the model is correct, but a greater possibility of concluding the model is incorrect when it is really correct (a *Type II error*). Larger values of α make it easier to conclude that the model is correct, but also increase the risk of accepting an incorrect model (a Type I error).

Regression Diagnostics

The regression diagnostic results display the values for the predicted values, residuals, and other diagnostic results.

Row This is the row number of the observation.

Predicted Values This is the value for the dependent variable predicted by the regression model for each observation.

Residuals These are the unweighted raw residuals, the difference between the predicted and observed values for the dependent variables.

Standardized Residuals The standardized residual is the raw residual divided by the standard error of the estimate $S_{y|x}$.

If the residuals are normally distributed about the regression, about 66% of the standardized residuals have values between −1 and +1, and about 95% of the standardized residuals have values between −2 and +2. A larger standardized residual indicates that the point is far from the regression. Values less than -2.5 or larger than 2.5 may indicate outlying cases.

Studentized Residuals The Studentized residual is a standardized residual that also takes into account the greater confidence of the predicted values of the dependent variable in the "middle" of the data set. By weighting the values of the residuals of the extreme data points (those with the lowest and highest independent variable values), the Studentized residual is more sensitive than the standardized residual in detecting outliers.

This residual is also known as the internally Studentized residual, because the standard error of the estimate is computed using all data.

Studentized Deleted Residuals The Studentized deleted residual, or externally Studentized residual, is a Studentized residual which uses the standard error of the estimate $S_{y|x(-i)}$, computed after deleting the data point associated with the residual. This reflects the greater effect of outlying points by deleting the data point from the variance computation.

The Studentized deleted residual is more sensitive than the Studentized residual in detecting outliers, since the Studentized deleted residual results in much larger values for outliers than the Studentized residual.

Figure 8–27
Regression Report

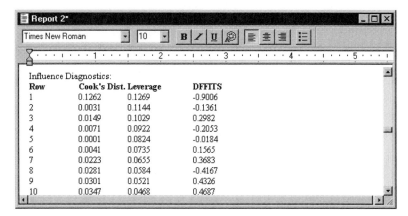

Row	Cook's Dist.	Leverage	DFFITS
1	0.1262	0.1269	-0.9006
2	0.0031	0.1144	-0.1361
3	0.0149	0.1029	0.2982
4	0.0071	0.0922	-0.2053
5	0.0001	0.0824	-0.0184
6	0.0041	0.0735	0.1565
7	0.0223	0.0655	0.3683
8	0.0281	0.0584	-0.4167
9	0.0301	0.0521	0.4326
10	0.0347	0.0468	0.4687

Influence Diagnostics

Row This is the row number of the observation.

Cook's Distance Cook's distance is a measure of how great an effect each point has on the estimates of the parameters in the regression equation. It is a measure of how much the values of the regression coefficients would change if that point is deleted from the analysis.

Values above 1 indicate that a point is possibly influential. Cook's distances exceeding 4 indicate that the point has a major effect on the values of the parameter estimates.

Leverage Leverage values identify *potentially* influential points. Observations with leverages a two times greater than the expected leverages are potentially influential points.

The expected leverage of a data point is $\frac{p}{n}$, where there are p parameters and n data points.

Because leverage is calculated using only the dependent variable, high leverage points tend to be at the extremes of the independent variables (large and small values),

Figure 8–28
Regression Report

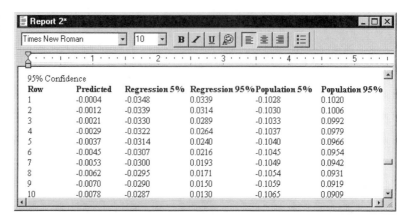

where small changes in the independent variables can have large effects on the predicted values of the dependent variable.

DFFITS The DFFITS$_i$ statistic is a measure of the influence of a data point on regression prediction. It is the number of estimated standard errors the predicted value for a data point changes when the observed value is removed from the data set before computing the regression coefficients.

Predicted values that change by more than 2.0 standard errors when the data point is removed are potentially influential.

95% Confidence Intervals

If the confidence interval does not include zero, you can conclude that the coefficient is not zero with the level of confidence specified. This can also be described as $P < \alpha$ (alpha), where α is the acceptable probability of incorrectly concluding that the coefficient is different than zero, and the confidence interval is $100(1 - \alpha)$.

The confidence level for both intervals is fixed at 95% (α=0.05).

Row This is the row number of the observation.

Predicted Values This is the value for the dependent variable predicted by the regression model for each observation.

Regression The confidence interval for the regression gives the range of variable values computed for the region containing the true relationship between the dependent and independent variables, for the specified level of confidence. The 5% values are lower limits and the 95% values are the upper limits.

Population The confidence interval for the population gives the range of variable values computed for the region containing the population from which the observations were drawn, for the specified level of confidence. The 5% values are lower limits and the 95% values are the upper limits.

Regression Equation Libraries and Notebooks

Regression equations are stored in notebook files just as other SigmaPlot documents. Notebooks that are used to organize and contain only regression equations are referred to as libraries, and distinguished from ordinary notebooks with a file extension of .JFL. These library notebooks can be opened and modified like any other notebook file. You can also use ordinary SigmaPlot notebooks (.JNB) as equation libraries, as well as save any notebook as a .JFL file.

Regression equations within notebooks are indicated with a ⌁ icon that appears next to the equation name.

The equations that appear in the Regression Wizard are read from a default regression library. The way the equations are named and organized in the equations panel is by using the section name as the category name, and the entry name as the equation name.

Figure 8–29
The Standard Regression
Equation Library

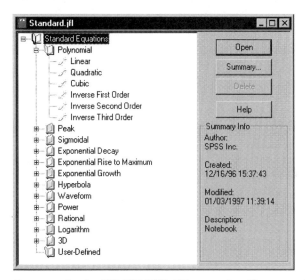

For example, the STANDARD.JFL regression library supplied with SigmaPlot has twelve categories of built-in equations:

➤ Polynomial

➤ Peak

➤ Sigmoidal

➤ Exponential Decay

➤ Exponential Rise to Maximum

➤ Exponential Growth

➤ Hyperbola

➤ Waveform

> ➤ Power
> ➤ Rational
> ➤ Logarithm
> ➤ 3D

These categories correspond to the section names within the STANDARD.JFL notebook. The equations for these different categories are listed in Appendix , "Regression Equation Library".

To see the library currently in use, click the Back button from the Regression Wizard equation panel. Previously selected libraries and open notebooks can be selected from the Library Drop Down list.

Opening an Equation Library

A regression equation library can be opened, viewed and modified as any ordinary notebook. To open a regression library:

➤ Click the Open toolbar button, select *.JFL as the file type from the File Type drop-down list, then select the library to open, or

➤ Click the Open button in the Regression Wizard library panel to open the current library. The library panel can be reached by clicking Back from the Equations panel.

You can copy, paste, rename and delete regression equations as any other notebook item. Opening a regression equation directly from a notebook automatically launches the Regression Wizard with the variables panel selected.

Using a Different Library for the Regression Wizard

You can also select another notebook or library as the source for the equations in the Regression Wizard. Selecting a different equation library changes the categories and equations listed Regression Wizard equations panel.

Figure 8–30
Selecting the Regression
Equation Library

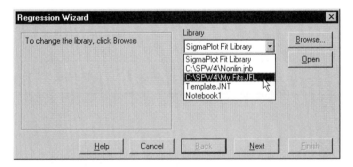

To change the library:

1. Start the Regression Wizard by pressing F5 or choosing the Statistics menu Regression Wizard command.

2. Click Back to view the library panel. To change the library used, enter the new

library path and name, or click Browse...

3. The File Open dialog box appears. Change the path and select the file to use as your regression library. When you start the Regression Wizard next, it will continue to use the equation library selected in the library panel.

Σ Note that opening a regression equation directly from a notebook *does not* reset the equation library.

Curve Fitting Date And Time Data

You can run the Regression wizard on data plotted versus calendar times and dates. Dates within and near the twentieth century are stored internally as very large numbers. However, you can convert these dates to relatively small numbers by setting Day Zero to the first date of your date, then converting the date data to numbers. After curve fitting the data, you can switch the numbers back to dates.

Figure 8–31
You can curve fit
dates, but you
must convert the
dates to
numbers first.

Time only data
(as shown in
column 1) does
not require a
conversion.

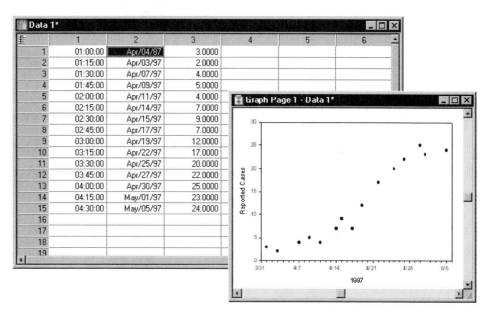

Σ If you have entered clock times only, then you can directly curve fit those time without having to convert these to numbers. Time only entries assume the internal start date of 4713 B.C. (the start of the Julian calendar). However, if you have entered times using a more recent calendar date, you must convert these times to numbers as well.

To convert your dates to numbers:

1. Choose the Tools menu Options command, then select Date and Time from the Show Settings For list.

Figure 8–32
Setting Day Zero

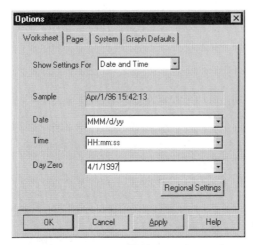

2. Set Day Zero to be the first date of your data, or to begin very close to the starting date of your data. You must include the year as well as month and day.

3. Click OK, then view the worksheet and select your data column. Choose the Format menu Cells command and choose Numeric.

Your dates are converted to numbers.

Figure 8–33
Converting Dates
to Numbers

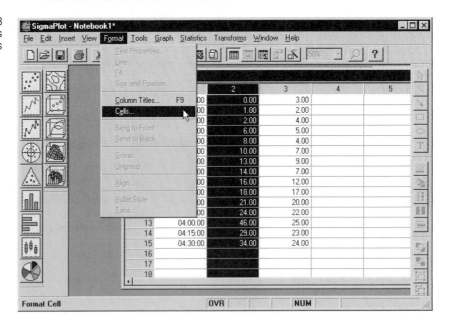

These numbers should be relatively small numbers. If the numbers are large, you did not select a Day Zero near your data starting date.

4. If the axis range of you graph is manual, convert it back to automatic. Select the axis, then open the Graph Properties dialog box and change the range to Automatic.

5. Click you curve and run your regression. When you are finished, you must convert the original and fitted curve x variable columns back to dates.

6. Select each column, then choose the Format menu Cells command and choose Date and Time, then click OK.

Figure 8–34
Selecting the Regression
Equation Library

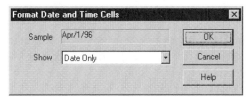

When the columns are converted back to dates, the graph should rescale and you have completed your date and time curve fit.

Figure 8–35
The Data and Fitted Curve X
Variables Converted Back to
Dates and Graphed

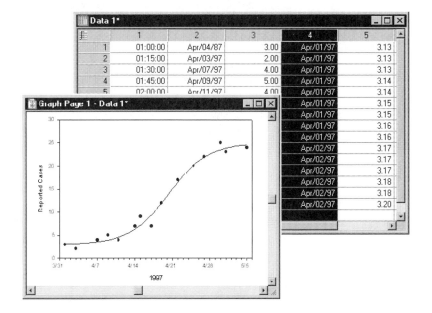

Regression Results Messages

When the initial results of a regression are displayed, a message about the completion status appears. Explanations of the different messages are found below.

Completion
Status Messages

Converged, tolerance satisfied. This message appears when the convergence criterion, which compares the relative change in the norm to the specified tolerance, is satisfied. Note that this result may still be false, caused by a local minimum in the sum of squares.

Converged, zero parameter changes. The changes in all parameters between the last two iterations are less than the computer's precision.

Did not converge, exceeded maximum number of iterations. More iterations were required to satisfy the convergence criteria. Select More Iterations to continue for the same number of iterations or increase the number of iterations specified in the Options dialog box and rerun the regression.

Did not converge, inner loop failure. There are two nested iterative loops in the Marquardt algorithm. This diagnostic occurs after 50 sequential iterations in the inner loop. The use of constraints may cause this to happen due to a lack of convergence. In some cases, the parameter values obtained with constraints are still valid, in the sense that they result in good estimates of the regression parameters.

Terminated by user. You pressed Esc, or selected the Cancel button and terminated the regression process.

Function overflow using initial parameter values. The regression iteration process could not get started since the first function evaluation resulted in a math error. For example, if you used $f = sqrt(-a*x)$, and the initial a value and all x values are positive, a math error occurs. Examine your equation, parameter values and independent variable values, and make the appropriate changes.

Parameters may not be valid. Array ill conditioned on final iteration. During the regression iteration process the inverse of an array (the product of the transpose of the Jacobian matrix with itself) is required. Sometimes this array is nearly singular (has a nearly zero determinant) for which very poor parameter estimates would be obtained.

SigmaPlot uses an estimate of the "condition" of the array (ill conditioned means nearly singular) to generate this message (see Dongarra, J.J., Bunch, J.R., Moler, C.B., and Stewart, G.W., *Linpack User's Guide*, SIAM, Philadelphia, 1979 for the computation of condition numbers).

Usually this message should be taken seriously, as something is usually very wrong. For example, if an exponential underflow has occurred for all x values, part of the equation is essentially eliminated. SigmaPlot still tries to estimate the parameters associated with this phantom part of the equation, which can result in invalid parameter estimates.

A minority of the time the "correct," though poorly conditioned, parameters are obtained. This situation may occur, for example, when fitting polynomial or other linear equations.

Parameters may not be valid. Array numerically singular on final iteration. This is a variant of the above condition. Instead of using the condition number the inverted array is multiplied by the original array and the resulting array elements are tested (the off diagonal elements are compared to 0.0 and the diagonal elements compared to 1.0).

If the absolute value of any off diagonal element or difference of the diagonal element from 1.0 is greater than a specified tolerance, then the original array is considered to be singular.

Parameters may not be valid. Overflow in partial derivatives. The partial derivatives of the function to be fit, with respect to the parameters, are computed numerically using first order differences.

Math errors from various sources can cause errors in this computation. For example if your model contains exponentials and the parameters and independent variable values cause exponential underflows, then the numerical computation of the partial derivative will be independent of the parameter(s). SigmaPlot checks for this independence.

Check the parameter values in the results screen, the range of the independent variable(s) and your equation to determine the problem.

Error Status Messages

Bad constraint. The regression cannot proceed because a constraint you defined either was not linear or contained syntax errors.

Invalid or missing 'fit to' statement. The regression lacks a fit to statement, or the fit to statement contains one or more syntax errors.

No observations to fit. The regression cannot proceed unless at least one x,y data pair (observation) is included. Check to be sure that the data columns referenced in the regression specifications contain data.

No parameters to fit. The regression specifications do not include any parameter definitions. To add parameter definitions, return to the Edit Regression dialog box and type the parameter definitions, in the Parameters edit window.

No weight statement. The regression specifications include a fit to statement with an unknown weight variable. Check the Variables edit window to see if a weight variable has been defined and that this corresponds to the variable in the regression statement.

Not enough or bad number of observations. In regression, the x and y data sets must be of the same size. The data sets (x and y columns) you specified contain unequal numbers of values.

Problem loading the file [Filename]. File too long; truncated. The fit file you tried to load is too long. Regression files can be up to 50 characters wide and 80 lines long. Any additional characters or lines were truncated when the file was loaded into the Edit Window.

Section has already been submitted. This regression section has already been defined.

Symbol [Variable or Function] has not been defined. The fit to statement in the regression definition contains an observed variable which is undefined, or the fit to statement in the regression definition contains an undefined function. Examine the regression specifications you have defined and be sure that the dependent variable listed in the regression statement exists and corresponds to the variable defined in the Variables edit window and that the function listed in the regression statement exists and corresponds to the function you defined in the Equations edit window.

Unreferenced variable. The regression specifications define a parameter that is not referenced in any other statements. Either delete the parameter definition, or reference it in another statement.

Editing Code

You can edit a regression equation by clicking the Edit Code button from the Regression Wizard. Regression equations can be selected from within the wizard, or opened from a notebook directly.

You can also create new regression equations. Creating a new equation requires entry of all the code necessary to perform a regression. This chapter covers:

➤ Selecting an equation for editing (see page 184)

➤ Entering equation code (see page 186)

➤ Defining constants (see page 188)

➤ Entering variables code (see page 192)

➤ Entering parameters code (see page 196)

➤ Entering code for parameter constraints and other options (see page 197)

About Regression Equations

Equations contain not only the regression model function, but other information needed by SigmaPlot to run a regression. All regression equations contain code defining the equations, parameter settings, variables, constraints, and other options used.

To edit the code for an equation, you need to either open and edit an existing equation, or create a new equation.

Protected Code for Built-in Equations   All built-in equations provided in STANDARD.JFL have protected portions of code which can be viewed and copied but not edited. However, you may use Add As to create a duplicate entry that can be edited, and you can also copy a built-in equation from the library to another notebook or section and edit it.

Using .FIT Files   For information on opening and editing SigmaPlot 3.0 and earlier .FIT files, see "Opening .FIT Files" on page 143.

Opening an Existing Equation

You can open an equation by:

➤ double-clicking an equation icon in a notebook window, or selecting the equation then clicking Open

➤ starting the Regression Wizard, then selecting the equation by category and name

Figure 9–1
Opening an Equation
from a Notebook

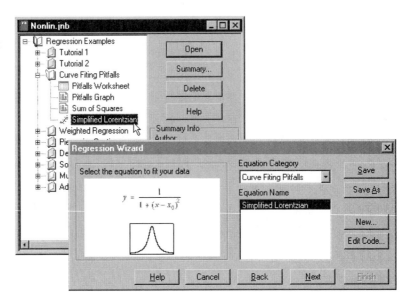

You can also double-click and equation in a notebook while the Regression Wizard is open to switch to that equation.

Once an equation is opened, you can edit it by clicking the Edit Code button.

Creating a New Equation

If you require an equation that does not appear in the standard equation library, you can create a new equation.

New equations can be created by:

➤ Clicking the New button in the Regression Wizard

➤ Choosing File menu New command, and selecting Regression Equation

➤ Right-clicking in the notebook window, and choosing New, Regression Equation from the shortcut menu

A new equation document has no default settings for the equations, parameters, variables, constraints, or other options.

To create a new equation from within the Regression Wizard:

1. Open the Regression Wizard by pressing F5 or by choosing the Statistics menu Regression Wizard command.

2. Click New to create a new equation document. The Regression dialog box appears.

Figure 9–2
Selecting Regression
Equation from
the New dialog box

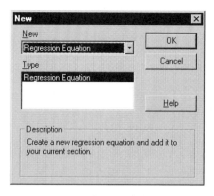

To use the File Menu New Command:

1. Select the notebook section where you want to add the equation. If you want the equation to be created in a new section, click the notebook icon.

2. Choose the File menu New command, and select Regression Equation from the New drop-down list.

3. Click OK to create the new equation. The Regression dialog box opens.

To create an Equation from the Notebook View:

1. Right-click the section where you want the equation to go. If you want the equation to be created in a new section, right-click the notebook icon.

2. Choose New from the shortcut menu, and choose Regression Equation. The Regression dialog box opens.

Figure 9–3
Creating a New Equation
from the Notebook

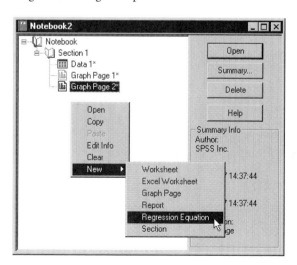

Copying Equations	You can copy an existing equation from any notebook view to another, and modify it as desired.
Adding Equations as New Entries	Equations can also be edited from within the Regression Wizard, and added as new equations to the current library using the Add As button in the Regression dialog box.

Entering Regression Equation Settings

To enter the settings for new equations, click the desired edit window in the Regression dialog box and enter your settings.

Figure 9–4
The Regression Dialog

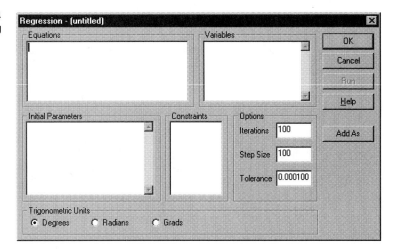

This section covers the minimum steps required to enter the code for a regression equation. For more information on entering the code for each section, see:

➤ "Equations" on page 191
➤ "Variables" on page 192
➤ "Weight Variables" on page 195
➤ "Initial Parameters" on page 196
➤ "Constraints" on page 197
➤ "Other Options" on page 198

Adding Comments	Comments are placed in the edit box by preceding them with an apostrophe ('), or a semicolon (;). You can also use apostrophes or semicolons to comment out equations instead of deleting them.

Entering Equations

To enter the code for the Equation section:

1. Click in the Equation window and type the regression equation model, using the transform language operators and functions.

   The equation should contain all of the variables you plan to use as independent variables, as well as the name for the *predicted* dependent variable (which is *not* your y variable). You can use any valid variable name for your equation variables and parameters, but short, single letter names are recommended for the sake of simplicity.

   Omit the observed dependent variable name from the regression model. The observed dependent variable (typically your y variable) is used in the *fit statement*.

2. Press the Enter key when finished with the regression equation model, then type the fit statement. The simplest form of the fit statement is:

   fit *f* to *y*

   Where *f* is the predicted dependent variable from the regression model, and *y* is the variable that will be defined as the observed dependent variable (typically the variable plotted as y in the worksheet).

3. You can also define whether or not weighting is used. For more information on how to perform weighted regressions, see "Weight Variables" on page 195.

Figure 9–5
Entering the Regression
Equation
and the
Regression Statement

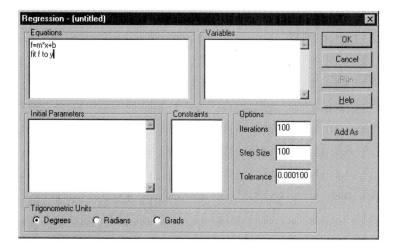

Example The code

f=m*x+b
fit f to y

can be used as the model for the function $f(x) = mx + b$, and also defines y as the observed dependent variable. In this example, x is the independent variable, and m and b the equation parameters.

Defining Constants

Constants that appear in the equations can also be defined under the equations heading. If you decide that an equation parameter should be a constant rather than a parameter to be determined by the regression, define the value for that constant here, then make sure you don't enter this value in the parameters section.

Constants defined here appear under the Constants option in the Regression Options dialog box.

Entering Variables

Independent, dependent, and weighting variables are defined in the Variables section. One of the variables defined must be the observed values of the dependent variable: that is, the "unknown" variable to be solved for. The rest are the independent variables (predictor, or known variables) and an optional weighting variable.

To define your variables:

1. Click in the Variables section and type the character or string you used for the first variable in your regression equation.

2. Type an equal sign (=), then enter a range for the variable. Ranges can be any transform language function that produces a range, but typically is simply a worksheet column.

    Note that the variable values used by the Regression Wizard depend entirely on what are selected from the graph or worksheet; the values entered here are only used if the From Code data format is selected, or if the regression is run directly from the Regression dialog box.

3. Repeat these steps for each variable in your equation. Up to ten independent variables can be defined, but you must define at least one variable for a regression equation to function. The curve fitter checks the variable definitions for errors and for consistency with the regression equation.

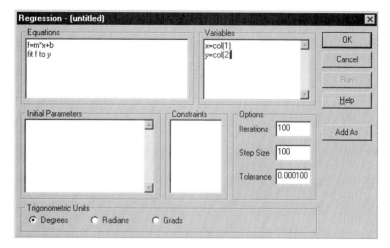

Figure 9–6
Entering the Variable
Definitions

Example: To define x and y as the variables for the equation code

f=m*x+b
fit f to y

you could enter the code

x=col(1)
y=col(2)

which defines an x variable as column 1 and a y variable as column 2, using these columns whenever the regression is run directly from the code.

Automatic Initial Parameter Estimation Functions

Any user-defined functions you plan on using to compute initial parameter estimates must be entered into the Variables section. For more information on how to code initial parameter estimate function, see "Automatic Determination of Initial Parameters" on page 200.

Entering Initial Parameters

Parameters are the equation coefficients and offset constants that you are trying to estimate in your equation model. The definitions or functions entered into the Parameters sections determine which variables are used as parameters in your equation model, and also their initial values for the curve fitter.

The curve fitter checks the parameter equations for errors and for consistency with the regression equations.

To enter initial parameter values:

1. Click in the Initial Parameters section and type the name of the first parameter as it appears in your equation model, followed by an equals (=) sign.

2. Enter the initial parameter value used by the curve fitter. Ideally, this should be as close to the real value as possible. This value can be numeric, or a function

that computes a "good guess" for the parameter. Using a function for the initial parameter value is called automatic parameter estimation. For more information on parameter estimation, see "Automatic Determination of Initial Parameters" on page 200.

Example If your data for the equation code

f=m*x+b
fit f to y

appear to rise to the right and run through the origin, you could define your initial parameter as

m=0.5
b=0

These are good initial guesses, since the *m* coefficient is the slope and the *b* constant is the y-intercept of a straight line.

Constraints   Parameter Constraints are completely optional, and should only be entered if you suspect they will improve the performance of the curve fitter. See "Constraints" on page 197 for when and how to enter constraints.

Options   The Iterations, Step Size and Tolerance options sometimes can be used to improve or limit your curve fit. The default settings work for the large majority of cases, so you do not need to change these setting unless truly required. For conditions that may call for the use of these options, see "Curve Fitting Pitfalls" on page 217. For more information on the effect of these options, see "Other Options" on page 198.

Saving Equations

Once you are satisfied with the settings you have entered into the Regression dialog box, you can save the equation. Clicking OK automatically updates the equation entry in the current notebook or regression library. If you created a new equation, you are prompted to name it before it is added to your notebook.

If you are editing an existing equation, you can click Add As to add the code as a new equation to the current library or notebook.

In order to save your changes to disk, you must also save the notebook or library. Changes to your current regression library are automatically saved when you close the wizard. You can also save changes before you close the wizard by clicking the Save button. Click Save As to save the regression library to a new file.

If your equation is part of a visible notebook, you can save changes by saving the notebook using the Save button or the File menu Save or Save As commands.

Note that when an equation is edited using the Regression Options dialog box, all the changes are also automatically updated and saved.

Saving Equation Copies with Results

You can save equations along with the targeted page or worksheet while saving your regression results. Just check the Add Equation to Notebook option in the results panel, and a copy of the equation used is added to the same section as reports and other results.

Equations

The Equation section of the Regression dialog box defines the model used to perform the regression as well as the names of the variables and parameters used.

The regression equation code is defined using the transform language operators and functions. The equation must contain all of the variables you wish to use. These include all independent variables, the *predicted* dependent variable, and observed dependent variable. All parameters and constants used are also defined here.

The Equation code consists of two required components:

➤ The *equation model* describing the function(s) to be fit to the data
➤ The *fit statement*, which defines the predicted dependent variable and, optionally, the name of a weighting variable

The independent variable and parameters are defined within the equation function. Also, any constants that are used must also be defined under the Equations section.

Defining the Equation Model

The equation model sets the predicted variable (called f in all built-in functions) to be a function of one or more independent variables (called x in the built-in two-dimensional Cartesian functions) and various unknown coefficients, called parameters.

The model may be described by more than one function. For example, the following three equations define a dependent variable f, which is a constant for $x < 1$ and a straight line for $x \geq 1$.

f = if (x < 1, constant (x), line (x))
constant (x) = c
line (x) = a + b $*$ x

Number of Parameters

You can enter and define up to 25 parameters, but a large number of parameters will slow down the regression process. You can determine if you are using too many parameters by examining the *parameter dependencies* of your regression results. Dependencies near 1.0 (0.999 for example) indicate that the equation is overparameterized, and that you can probably remove one or more dependent parameters. For more information on parameter dependencies, see "Interpreting Initial Results" on page 163.

Defining the Fit Statement

The most general form of the fit statement is:

fit *f* to *y* with weight *w*

f identifies the predicted dependent variable to be fit to the data in the set of equations, as defined by the model.

y is the observed dependent variable, later defined in the Variables section, whose value is generally determined from a worksheet column.

w is the optional weight variable, also defined in the Variables section. Any valid variable name can be used in place of f, y, and w.

If the optional weighting variable is not used, the fit statement has the form:

fit f to y

Defining Constants

Constants are simply defined by setting one of the parameters of the equation model to a value, using the form

constant=value

For example, one commonly used constant is pi, defined as

pi=3.14159265359

Defining Alternate Fit Statements

You can create alternate fit statements that call different weight variables. These statements appear as fit statements preceeded by two single quotes (", not a double quote).

For each weight variable you define, you can create a weighting option by adding commented fit statements to the equation window.

For example, an Equation window that reads

f=a*exp(-b*x)+c*exp(-d*x)+g*exp(-h*x)
fit f to y
"fit f to y with weight Reciprocal

would display the option Reciprocal in the Regressions Options dialog box Fit With Weight list.

Variables

Independent, dependent, and weighting variables are defined in the Variables edit window. One of the variables defined must be the observed values of the dependent

Figure 9–7
An Equations Window with
Alternate Fit Statements

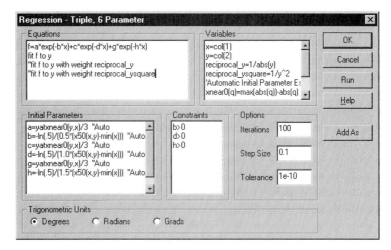

variable: that is, the "unknown" variable to be solved for. The rest are the independent variables (predictor, or known variables) and any optional weighting variables. Up to ten independent variables can be defined.

To define your variables, select the Variables edit window, then type the variable definitions. You generally need to define at least two variables—one for the dependent variable data, and at least one for the independent variable data.

Variable Definitions

Variable definitions use the form:

variable = *range*

You can use any valid variable name, but short, single letter names are recommended for the sake of simplicity (for example, x and y). The range can either be the column number for the data associated with each variable, or a manually entered range.

Most typically, the range is data read from a worksheet. The curve fitter uses SigmaPlot's transform language, so the notation for a column number is:

col(*column,top,bottom*)

The *column* argument determines the column number or title. To use a column title for the column argument, enclose the column title in quotation marks. The *top* and *bottom* arguments specify the first and last row numbers and can be omitted. The default row numbers are 1 and the end of the column, respectively. If both are omitted, the entire column is used. For example, to define the variable x to be column 1, enter:

x = col(1)

Data may also be entered directly in the variables section. For example, you can define y and z variables by entering:

$y = \{1,2,4,8,16,32,64\}$
$z = data(1,100)$

This method can have some advantages. For example, in the example above the data function was used to automatically generate Z values of 1 through 100, which is simpler than typing the numbers into the worksheet.

Σ Note that the Regression Wizard generally ignores the default variable settings, although it requires valid variable definitions in order to evaluate an equation. Variables are redefined when the variables are selected from within the wizard. However, you can force the use of the hard-coded variable definitions, either by selecting From Code as the data source, or running the regression directly from the Regression dialog box.

Transform Language Operations
You can use any transform language operator or function when defining a variable. For example:

$x = 10\wedge data(-2, \log(10.8),0.5)$
$y = ((col(2)-col(2)*(.277*col(1))\wedge 0.8))*1.0e-12$
$z = 1/sqrt(abs(col(3)))$

are all valid variable names.

User-Defined Functions
Any user-defined functions that are used later in the regression code must be defined in the Variables section.

Concatenating Columns
Constructor notation can be used to concatenate data sets. For example, you may want to fit an equation simultaneously to multiple y columns paired with one x column. If the x data is in column 1 and the y data is in columns 2 through 6, you can enter the following variable statements

$x = \{col (1), col (1), col (1), col (1), col (1)\}$
$y = \{col (2), col (3), col (4), col (5), col (6)\}$

The variable x is then column 1 concatenated with itself four times, and variable y is the concatenation of columns 2 through 6.

If the function to be fit is f, then the fit statement

fit f to y

fits f to the dependent variable values in columns 2 through 6 for the independent variable values in column 1.

Weight Variables

Variables used to perform weighted regressions are known as weight variables. All weight variables must be defined along with other variables in the Variables window.

Specifying the Weight Variable to Use

The use of weighting is specified by the Equation section code, which can call weight variables defined under Variables. Weight variables are selected from the fit statement, using the syntax

fit f to y with weight *w*

where *w* is the weight variable defined under Variables. See "Equations" on page 191 for additional details on how to define the fit statement.

Generally, a weight variable is defined as the reciprocal of either the observed dependent variable or its square. For example, if *y*=col(2) is the observed dependent variable, the weighting variable can defined as 1/col(2) or as 1/col(2)^2.

For a demonstration of weighting variable use, see "Example 2: Weighted Regression" on page 223.

Defining Optional Weight Variables

You can define more than one possible weight variable, and select the one to use from the Regression Options dialog box. Simply create multiple weight variables, then create alternate fit statement entries selecting the different weight variables in the Equations window. For more information on creating alternate fit statements, see "Defining Alternate Fit Statements" on page 192.

When to Use Weighting

Least squares regressions assumes that the errors at all data points are equal. When the error variance is not homogeneous, weighting should be used. If variability increases with the dependent variable value, larger dependent variable values will have larger residuals. Large residuals will cause the squared residuals for large dependent variable values to overwhelm the small dependent variable value residuals. The total sum of squares will be sensitive only to the large dependent variable values, leading to an incorrect regression.

You may also need to weight the regression when there is a requirement for the curve to pass through some point. For example, the (0,0) data point can be heavily weighted to force the curve through the origin.

Σ Note that if you use weighted least squares, the regression values are valid, but the statistical values produced for the curve are not.

The Weighting Process: Norm and Residuals Changes

The weight values are proportional to the reciprocals of the variances of the dependent variable. Weighting multiplies the corresponding squared term in the sum of squares, dividing the absolute value of the residual by its standard error. This causes all terms of the sum of squares to have a similar contribution, resulting in an improved regression.

For weighted least squares, the weights *w* are included in the sum of squares to be minimized.

$$SS = \sum_{i=1}^{n} w_i(y_i - \hat{y}_i)^2$$

When weighting is used, the norm that is computed and displayed in the Progress dialog box and initial results is $\sqrt{SS}$, and includes the effect of weighting. The residuals computed are the weighted residuals $\sqrt{w_i}(y_i - \hat{y}_i)$.

Initial Parameters

The code under the Initial Parameters section specify which equation coefficients and constants to vary and also set the initial parameter values for the regression.

To enter parameters, select the Initial Parameters window, then type the parameters definitions using the form:

parameter=initial value

All parameters must appear in the equation model. All equation unknowns not defined as variables or constants must be defined in Initial Parameters.

Initial Parameter Values

For the initial values, a "best guess" may speed up the regression process. If your equation is relatively simple (only two or three parameters), the initial parameter values may not be important. For more complex equations, however, good initial parameter values can be critical for a successful convergence to a solution.

Automatic Parameter Estimation

All built-in equations use a technique called *automatic parameter estimation*, which computes an approximation of the function parameters by analyzing the raw data. You can indicate the parameter value you wish to appear as the Automatic setting by typing two single quotes followed by the string Auto after the parameter setting. For example, entering the parameter line

a=max(y) ''Auto

tells the Regression Options dialog box to use *max(y)* as the Automatic parameter value for *a*.

This technique is further described under "Automatic Determination of Initial Parameters" on page 200.

Alternate Parameter Values

You can insert alternate parameter values that appear in the Regression Options dialog box Initial Parameter Values drop-down lists. To add an alternate, insert a new line after the default value, then type two single quotes, followed by the alternate parameter setting. For example, the line

d=-F(0)[2] ''Auto
''d=0.01

causes an alternate value of 0.01 to appear in the Regression Options dialog box Initial Parameter Values drop-down list for *d*.

Alternate parameter values are automatically inserted when different parameter values are entered into the Regression Options dialog box.

Constraints

Linear parameter constraints are defined under the Constraints section. A maximum of 25 constraints can be entered. Use of constraints is optional.

Constraints are used to set limits and conditions for parameter values, restricting the regression search range and improving regression speed and accuracy. Liberal use of constraints in problems which have a relatively large number of parameters is a convenient way to guide the regression and avoid searching in unrealistic regions of parameter space.

Valid Constraints

A constraint must be a linear equation of the parameters using an equality ($=$) or inequality ($<$ or $>$). For example, the following constraints for the parameters a, b, c, d, and e are valid:

a<1
10$_*$b+c/20 > 2
d−e = 15
a>b+c+d+e
whereas

a$_*$x<1

is illegal since x is not a constant, and

b+c^2>4
d$_*$e=1

are illegal because they are nonlinear.

Σ Although the curve fitter checks the constraints for consistency, you should still examine your constraint definitions before executing the regression. For example, the two constraints:

Figure 9–8
Entering Iteration, Step Size,
and Tolerance Options

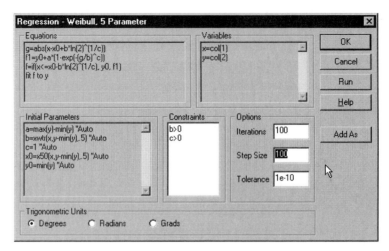

a<1
a>2

are inconsistent. The parameter *a* cannot be both less than 1 and greater than 2. If you execute a regression with inconsistent constraints, a message appears in the Results dialog box warning you to check your constraint equations.

Other Options

You can use several special options to influence regression operation. The different options can be used to speed up or improve the regression process, but their use is optional. The three options are:

➤ *Iterations*, the maximum number of repeated regression attempts.

➤ *Step Size*, the limit of the initial change in parameter values used by the regression as it tries different parameter values.

➤ *Tolerance*, one of the conditions that must be met to end the regression process. When the absolute value of the difference between the norm of the residuals from one iteration to the next is less than the tolerance, this condition is satisfied and the regression considered to be complete.

Options are entered in the Options section edit boxes. The default values are displayed for new equations. These settings will work for most cases, but can be changed to overcome any problems encountered with the regression, or to perform other tasks, such as evaluating parameter estimation.

Iterations

Setting the number of iterations, or the maximum number of repeated regression attempts, is useful if you do not want to regression to proceed beyond a certain number of iterations, or if the regression exceeds the default number of iterations.

The default iteration value is 100. To change the number of iterations, simply enter the maximum number of iterations in the Iterations edit box.

Evaluating Parameter Values Using 0 Iterations

Iterations must be non-negative. However, the setting Iterations to 0 causes no iterations occur; instead, the regression evaluates the function at all values of the independent variables using the parameter values entered under the Initial Parameters section and returns the results.

If you are trying to evaluate the effectiveness of automatic parameter estimation function, setting Iterations to 0 allows you to view what initial parameter values were computed by your algorithms.

Using zero iterations can be very useful for evaluating the effect of changes in parameter values. For example, once you have determined the parameters using the regression, you can enter these values plus or minus a percentage, run the regression with zero iterations, then graph the function results to view the effect of the parameter changes.

Step Size

The initial step size used by the Marquardt-Levenberg algorithm is controlled by the Step Size option. The value of the Step Size option is only indirectly related to changes in the parameters, so only relative changes to the step size value are important.

The default step size value is 100. To change the step size value, type a new value into the edit box. The step size number equals the largest step size allowed when changing parameter values. Changing the step size to a much smaller number can be used to prevent the curve fitter from taking large initial steps when searching around suspected minima.

For an example of the possible effects of step size, see "Curve Fitting Pitfalls" on page 217.

If you are familiar with this algorithm, step size is the inverse of the Marquardt parameter.

Tolerance

The Tolerance option controls the conditions that must be met in order to end the regression process. When the absolute value of the difference between the norm of the residuals from one iteration to the next is less than the tolerance, the regression is considered to be complete.

The curve fitter uses two stopping criteria:

➤ When the absolute value of the difference between the norm of the residuals (square root of the sum of squares of the residuals), from one iteration to the next, is less than the tolerance value, the iteration stops.

➤ When all parameter values stop changing in all significant places, the regression stops.

When the tolerance condition has been met, a minimum has usually been found.

The default value for tolerance is 0.0001. To change the tolerance value, type the required value in the Tolerance edit box. The tolerance number sets the value that must be met to end the iterations.

More precise parameter values can be obtained by decreasing the tolerance value. If there is a sharp sum of squares response surface near the minimum, then decreasing the tolerance from the default value will have little effect.

However, if the response surface is shallow about the minimum (indicating a large variability for one or more of the parameters), then decreasing tolerance can result in large changes to parameter values.

For an example of the possible effects of tolerance, see "Curve Fitting Pitfalls" on page 217.

Automatic Determination of Initial Parameters

SigmaPlot automatically obtains estimates of the initial parameter values for all built-in equations found in STANDARD.JFL. When automatic parameter estimation is used, you no longer have to enter static values for parameters yourself—the parameters determine their own values by analyzing the data.

Σ Note that it is only important that the initial parameter values are robust among varying data sets, i.e., that in most cases the curve fitter converges to the correct solution. The estimated parameters only have to be a "best guess" (somewhere in the same ballpark as the real values, but not right next to them).

You can create your own methods of parameter determination using the new transform function provided just for this purpose.

The general procedure is to smooth the data, if required, and then use functions specific to each equation to obtain the initial parameter estimates.

Consider the logistic function as an example. This function has the stretched "s" shape that changes gradually from a low value to a high value or vice versa.

The three parameters for this function determine the high value (a), the x value at which the function is 50% of the function's amplitude (x0) and the width of the transition (b). As expressed in the transform language, the function is entered into the Equation window as

f=a/(1+exp(-(x-x0)/b))
fit f to y

Noise in the data can lead to significant errors in the estimates of x0 and b. Therefore, a smoothing algorithm is used to reduce the noise in the data and three functions are then used on the smoothed data to obtain the parameter estimates.

To estimate the parameter a the maximum y value is used. The x value at 50% of the amplitude is used to estimate x0, and the difference between the x values at 75% and 25% of the amplitude is used to estimate b. As entered into the Initial Parameters window, these are

a=max(y) ''Auto
b=xwtr(x,y,.5)/4 ''Auto
x0=x50(x,y,.5) ''Auto

Both the fwhm and xwtr transform functions have been specifically designed to aid the estimation of function parameters. For more details on these specialized transform functions, see "Curve Fitting Functions" on page 25.

The ''Auto comment that follows each parameter is used to identify that parameter value as the Automatic setting from within the Regression Options dialog box.

Note that these values may not at all reflect the final values, but they are approximate enough to prevent the curve fitter from finding false or invalid results.

Regression Lessons

Lesson 1: Linear Curve Fit

This tutorial lesson is designed to familiarize you with regression fundamentals. The sample graph and worksheet files for the tutorials are located in the NONLIN.JNB Regression Examples notebook provided with SigmaPlot.

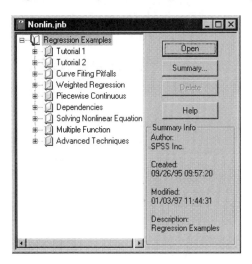

Figure 10–1
The NONLIN.JNB Notebook

In this lesson, you will fit a straight line to existing data points.

1. Open the Tutorial 1 Graph in the NONLIN.JNB notebook and examine the

graph. The points appear to nearly follow a straight line.

Figure 10–2
The Graph with
Unfitted Data Points

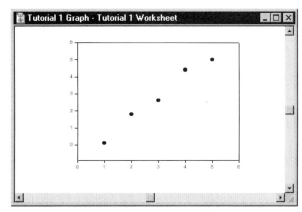

2. Choose the Statistics menu Regression Wizard command or press F5. The Regression Wizard dialog box displays lists of equations by category. If the Linear equation is not already selected, select the Polynomial category and select Linear as the equation name.

Figure 10–3
Selecting an Equation in the
Regression Wizard

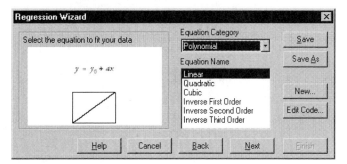

3. Click Next to proceed. The next panel prompts you to pick your x, or independent variable. Click the curve on the page to select it. Note that clicking the curve selects both the x and y variables for you.

Figure 10–4
Selecting the
Variables to Fit

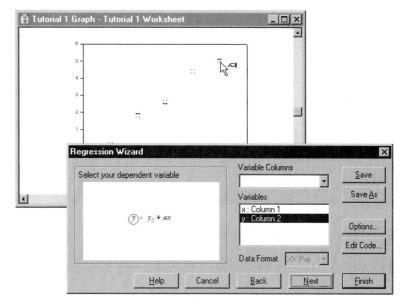

4. Click Next. The Iterations dialog box appears, displaying the progress of the fitting process. When the process is completed, the initial regression results are displayed.

5. Examine the results. The first result column is the parameter values; the intercept is −.94 and the slope is 1.24.

Figure 10–5
Examining Initial Results

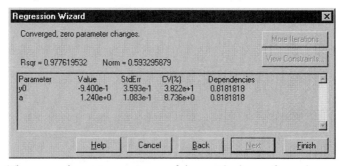

The next column is an estimate of the standard error for each parameter. The intercept has a standard error of about 0.36—not that good—and the slope has a standard error of about 0.10, which isn't bad.

The third column is the coefficient of variation (CV%) for each parameter. This is defined as the standard error divided by the parameter value, expressed as a percentage. The CV% for the intercept is about 38.2%, which is large in comparison to the CV% for the slope (about 8.7%).

The dependencies are shown in the last column. If these numbers are very close to 1.0, they indicate a dependency between two or more parameters, and you can probably remove one of them from your model.

Adding a Parameter Constraint

To make y always positive when x is positive, you cannot have a negative y intercept. You can recalculate the regression with this condition by constraining the parameter y_0 to be positive. That way y will never be negative when $x>0$.

1. From the initial results panel, click Back. The variables panel is displayed.

2. Click the Options button. The Options dialog box is displayed. Enter a value of

   y0>0

   into the Constraints edit box. This defines the constraint $y_0>0$, which forces the y intercept to be positive.

Figure 10–6
Adding a Parameter
Constraint Option to the
Regression Options
dialog box

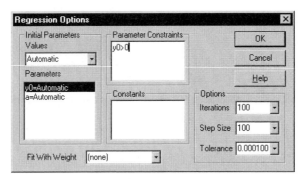

3. Click OK, then click Next to refit the data with a straight line, this time subject to the constraint $y_0>0$. When the initial results are displayed, the value for y_0 is now about 9.3×10^{-9}, very close to zero, and the slope has slightly decreased to a value of approximately 0.98.

Figure 10–7
The Results
of the Second Fit

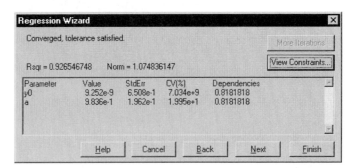

4. Select Constraints; the Constraints dialog box appears with the constraint y0>0 flagged with the label "(active)" indicating that it was used in the nonlinear regression.

Figure 10–8
The Constraints dialog box

Note that nonlinear regressions may find parameters that satisfy the constraints without having to activate some or all of the constraints. Constraints that are not used are not flagged as (active).

5. Click OK to return to the Nonlinear Regression Results dialog box, then click Next to proceed.

Saving Results

6. You can select the results to save for a regression. These results are destroyed by default each time you run another regression equation.

You can save some of your results to a worksheet, and other results to a text report. To save worksheet results, make sure the results you want saved are checked in the results list. You have the option to save parameter values, predicted dependent (y) variable values for the original independent (x) variable, and the residuals about the regression for each original dependent variable.

Figure 10–9
Saving the Nonlinear
Regression Results
Using the Keep Regression
Results dialog box

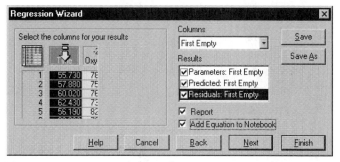

7. To save a text report, make sure the Report option is checked. The report for a nonlinear regression lists all the settings entered into the nonlinear regression dialog box, a table of the values and statistics for the regression parameters, and some regression diagnostics.

You can also save a copy of the regression equation you used to the same section as the page or worksheet on which you ran the regression. Check the Add Equation to Notebook option to save a copy of your equation.

Click Next to proceed.

Graphing Results

8. To plot the regression function on the existing graph, make sure the Add curve to Graph #1 option is checked.

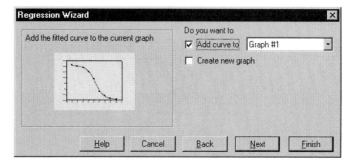

Figure 10–10
The Graph Results
Panel

9. Click Finish display your report and graphed results.

Figure 10–11
Regression
Results for a
Linear
Regression

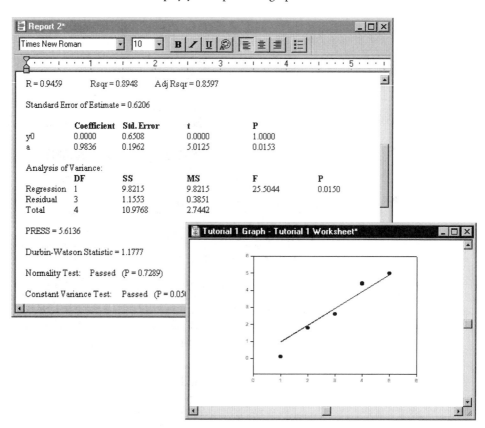

R = 0.9459   Rsqr = 0.8948   Adj Rsqr = 0.8597

Standard Error of Estimate = 0.6206

	Coefficient	Std. Error	t	P
y0	0.0000	0.6508	0.0000	1.0000
a	0.9836	0.1962	5.0125	0.0153

Analysis of Variance:

	DF	SS	MS	F	P
Regression	1	9.8215	9.8215	25.5044	0.0150
Residual	3	1.1553	0.3851		
Total	4	10.9768	2.7442		

PRESS = 5.6136

Durbin-Watson Statistic = 1.1777

Normality Test: Passed   (P = 0.7289)

Constant Variance Test: Passed   (P = 0.05(

Comparing Regression Wizard Results with Linear Regression Results

The original data for this graph could have been fitted automatically in SigmaPlot with a linear regression using the Statistics menu Linear Regression command. However, because you cannot specify constraints for the regression coefficients, a first order regression gives different results.

To add a linear regression to your original data plot:

1. Select the plot of your original data by clicking it on the graph, then choose the Statistics menu Linear Regression command.

Figure 10–12
Selecting a Linear
Regression

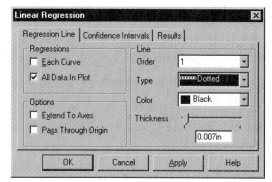

2. Select to draw a 1st order regression and pick a dotted line type for the regression line.

3. Click OK to accept the regression settings, then view the graph. Note the difference between the regression and the fitted line (use the View menu to zoom in on the graph if necessary).

Figure 10–13
Comparing the Fitted
Curve with a First
Order Regression

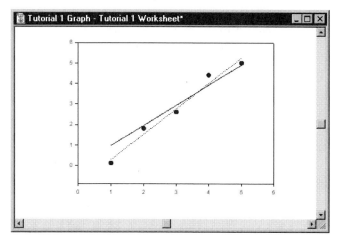

Note that if you had not used a parameter constraint, the result of the nonlinear regression would have been identical to the linear regression.

If desired, you can now save the graph and worksheet to a file using the File menu Save As command.

Lesson 2: Sigmoidal Function Fit

This tutorial leads you through the steps involved in solving a typical nonlinear function for a "real world" scenario.

Examining and Analyzing the Data

The data used for this tutorial represents blood pressure measurements made in the neck (carotid sinus pressure), and near the outlet of the heart (the mean arterial pressure).

These pressures are inversely related. If the blood pressure in your neck goes down, your heart needs to pump harder to provide blood flow to your brain. Without this immediate compensation, you could pass out every time you stood up.

Sensors in your neck detect changes in blood pressure, sending feedback signals to the heart. For example, when you first get out of bed in the morning, your blood tends to drain down toward your legs. This decreases the blood pressure in your neck, so the sensors tell the heart to pump harder, preventing a decrease in blood flow to the brain.

You can do an interesting experiment to demonstrate this effect. Stand up and relax for a minute, then take your pulse rate. Count the number of pulses in 30 seconds, then lie down and immediately take your pulse rate again. Your pulse rate will decrease as much as 25%. (Your heart doesn't have to pump as hard to get blood to the brain when you are lying down.)

1. Open the Tutorial 2 graph file by double-clicking the graph page icon in the Tutorial 2 section in the NONLIN.JNB notebook. Examine the graph. The two pressures are clearly inversely related. As one rises, the other decreases. The shape appears to be a reverse sigmoid, suggesting the use of a sigmoidal equation.

    A sigmoid shaped curve looks like an *S* that has had its upper right and lower left corners stretched. In this case, the *S* is backwards, since it starts at a large value, then decreases to a smaller value.

Various forms of the sigmoid function are commonly used to describe sigmoids.

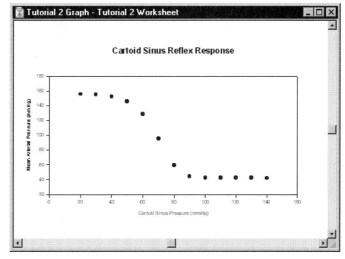

In this case, you will use the four parameter sigmoid function provided in the standard regression library.

2. Right-click the curve and choose Fit Curve. The Regression Wizard appears.

Figure 10–15
Choosing Fit Curve

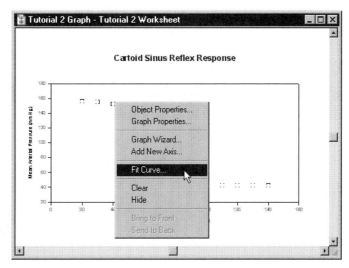

Select Sigmoidal as your equation category, and Sigmoid, 4 Parameter as your equation.

Figure 10–16
Selecting the Sigmoid, 4
Parameter Equation

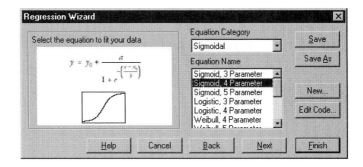

3. Click Next twice. If you have correctly selected the curve, the Iterations dialog box appears, displaying the value for each parameter and the norm for each iteration.

   Note that the iterations proceed more slowly than those for the linear fit. This is because the equation is much more complex and there are more parameters. Watch the *norm* value decrease—this number is an index of the fit closeness, and decreases as the fit improves.

   When the fit condition is satisfied, the initial results are displayed.

4. Examine the results. The first column displays the parameter value, and the next column displays the estimated standard error. The third column is the coefficient of variation (CV%) for each parameter. (Note that these CV% values are unrealistically good—the largest is about 3.9%. Generally, CV% values for physiological measurements are greater than 5%.)

Figure 10–17
The Fit Results for the Four
Parameter Sigmoid Function

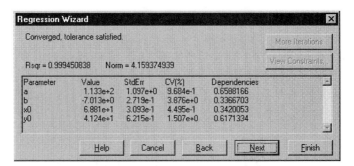

True nonlinear regression problems (like this sigmoidal fit, but unlike a linear fit) have CV% values that are not absolutely correct. However, they still can be used to compare the relative variability of parameters. For example, b (3.9) is more than eight times as variable as c (0.45).

None of the dependencies shown in the last column are close to 1.0, suggesting that the model is not over-parameterized.

5. To save the regression results and graph the curve, click Finish. A report along with worksheet data and a fitted curve are added to your notebook, worksheet and graph.

Figure 10–18
The Fitted Curve for the
Sigmoidal Data

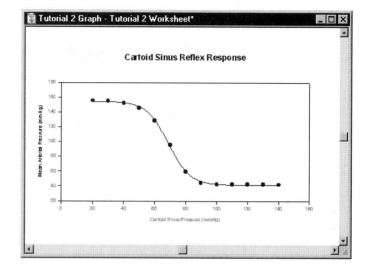

Fitting with A Different Equation

More than a single regression can be run and plotted on a graph. Typically, this is done to gauge the effects of changes to parameter values, or to compare the effect of a different fit equation.

In this case, try a five parameter logistic function instead of the four parameter version.

1. Press F5. The Regression Wizard appears. Select Sigmoid, 5 Parameter as your equation.

2. Click Next, then click the Options button. Enter a value of 5 into the Iterations box.

The iterations option specifies the maximum number of iterations to perform before displaying the current results. You can see if a long regression is working correctly by limiting the number of iterations to perform. If the regression does not complete within the number of iterations specified, you can continue by clicking the More Iterations option in the initial results panel. Click OK.

Figure 10–19
The Regression
Options dialog box

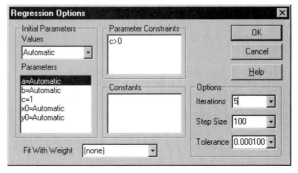

3. Click Next to calculate the new fit. Note that the Iteration dialog box now says "Iteration n of 5." Each iteration also requires much more time to calculate, and more iterations are required to produce a result.

After five iterations, the initial results panel is displayed. Note that the More Iterations option is no longer dimmed. Click More Iterations for five more iterations.

Figure 10–20
Results of the Five Parameter
Logistic Equation Fit After
Five Iterations

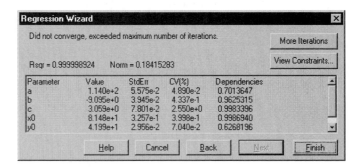

The regression continues to completion, converging after four more iterations.

Figure 10–21
Results of the Five
Parameter Logistic
Equation Fit More
Iterations

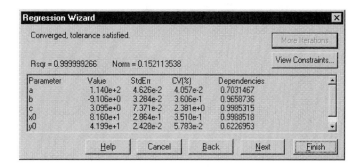

4. Examine the results. The norm value, standard deviations and CV% values are smaller than for the four parameter fit, indicating that this may be a better fit. However, two of the dependencies are close to 1.0, suggesting that the fifth parameter may not have been needed.

5. Click Finish to save the results of this regression. Another report, more data, and the curve for this regression equation are all added to your notebook.

To distinguish between the two regression lines, double-click one of them and change the line color to blue, then use the Plot drop-down list to change the other regression curve to red.

Compare the curve fits visually. As expected, the five parameter function appears to fit slightly better.

Figure 10–22
The SigmaPlot Graph with
both Four and Five Parameter
Logistic Equation Fit Results

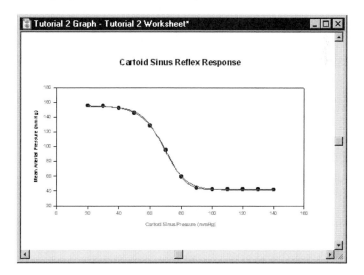

Notes

11 Advanced Regression Examples

Curve Fitting Pitfalls

This example demonstrates some of the problems that can be encountered during nonlinear regression fits.

Peaks in chromatograph data are sometimes fit with sums of Gaussian or Lorentzian distributions. A simplified form of the Lorentzian distribution is:

$$y = \frac{1}{1 + (x - x_0)^2} \qquad -\infty < x < \infty$$

where x_0 is the location of the peak value.

A graph of the distribution for $x_0 = 0$ is shown in Figure 11–1.

1. Open the Pitfalls worksheet and graph by double-clicking the Pitfalls Graph in the NONLIN.JNB notebook. Note the positions of data points on the curve.

2. Open the Simplified Lorentzian regression equation by double-clicking it in the Regression Examples notebook. The Regression Wizard opens and displays the variables panel (see Figure 11–7 on page 220).

3. Click one of the symbols on the graph so that the Variables selected are Columns 1 and 2.

   The object is to determine the peak location x_0 for the data. Since this data was generated from the Lorentzian function above using $x_0 = 0$, the regression should always find the parameter value $x_0 = 0$.

How the Curve Fitter Finds x_0

To find x_0, the curve fitter computes the sum of squares function:

$$\sum_{i=1}^{3} \left[f(x_i) - y_i \right]^2$$

as a function of the parameter x_0. The graph of this result using the x and y data is provided in Figure 11–1. The curve fitter then searches this parameter space for any x_0 value where a relative minimum exists.

The sum of squares for x_0 has two minima—an absolute minimum at $x_0 = 0$ and a relative minimum at $x_0 = 4.03$—and a maximum at 2.5. As the curve fitter searches for a minimum, it may stumble upon the local minimum and return an incorrect result. If you start exactly at a maximum, the curve fitter may also remain there.

Figure 11–1
The Plot of the Sum of Squares for $x_0 = 0$ of a Simplified Lorentzian Distribution

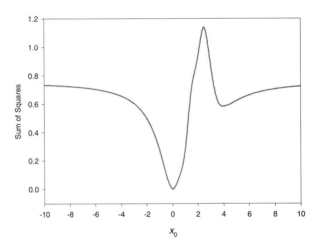

4. **False convergence caused by a small step size**   Click the Options button. Note that the value of x0 is set to 1000, and the Step Size option is set to 0.000001.

Figure 11–2
The Regression Options dialog box Showing Step Size Set to 0.00001

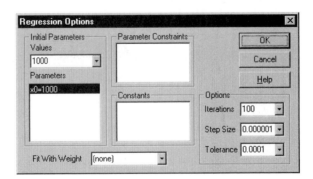

Click OK, then click Next.

Using the large initial value of x_0 and a small step size, the curve fitter takes one small step, finds that there is no change in the sum of squares using the default value for tolerance (0.0001), and declares the tolerance condition is satisfied. The very low slope in the sum of squares at this large x_0 value causes the regression to stop.

Figure 11–3
The Results Using
a Step Size of 0.00001

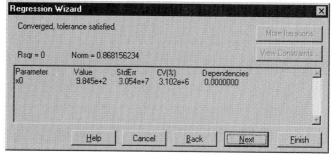

5. **False convergence caused by a large step size and tolerance** Click Back, then click the Options button. Open the Step Size list and select 100; this is the default step size value.

Figure 11–4
Selecting a Step Size of 100

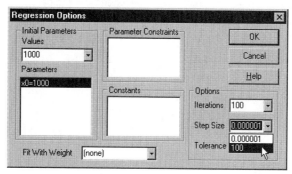

6. Click OK, then click Next. The curve fitter takes a large step, reaches negative x_0 values, and finds a value $x_0 = -546$ for which the tolerance is satisfied.

Figure 11–5
The Results
Using a Step Size of 100

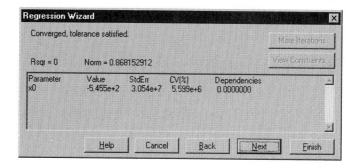

The sum of squares function asymptotically approaches the same value for both large positive and negative values of x, so the difference of the sum of squares for $x_0 = 1000$ and $x_0 = -546$ is within the default value for the tolerance.

7. **Reducing tolerance for a successful convergence**   Click Back, then click Options again. Change the Tolerance value to 0.000001, then click OK.

Figure 11–6
Changing the Tolerance to
0.0001

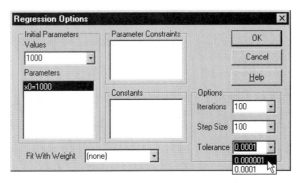

8. Click Next. The regression continues beyond $x_0 = -546$ and successfully finds the absolute minimum at $x_0 = 0$.

Figure 11–7
The Results of Using a Step
Size of 100
and Tolerance of 0.000001

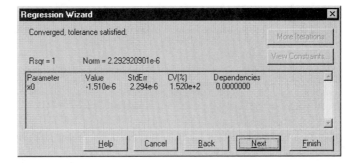

Summary

When you used a poor initial parameter value, you needed to use a large initial step size to get the regression started, and you had to decrease the tolerance to keep the regression from stopping prematurely. Poor initial parameters can arise also when using the Automatic method of determining initial parameters as well as when constant values are used.

You will now use initial parameter values which result in convergence to a local minimum and a local maximum.

1. **Finding a local minimum**   Click Back, then click the Options button. Change the initial value of x_0 to 10 using the drop down Parameter Values list.

Figure 11–8
Changing the Initial
Parameter Value of x_0 to 10
and the Tolerance to 0.0001

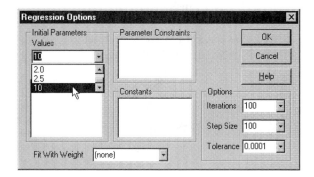

2. Select the Tolerance option and change the tolerance back to the default value of 0.0001, then click OK.

3. Click Next. The regression converges to $x_0 = 4.03$, which corresponds to the local minimum.

Figure 11–9
The Nonlinear Regression
Results Using an Initial
Parameter Value of $x_0 = 10$

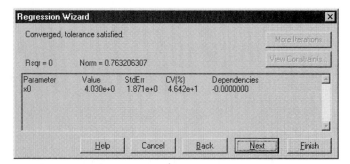

In this example, you know that a local minimum was found by viewing the sum of squares function for the single parameter x_0. However, when there are many parameters, it is usually not obvious whether an absolute minimum or a local minimum has been found.

4. **Finding a local maximum** Click Back, then click the Options button. Change the initial parameter value of x_0 to 2.5, then click OK.

5. Click Next. Because this initial parameter value happens to correspond to the maximum of the sum of squares function, the regression stops immediately. The slope is zero within the default tolerance, so the curve fitter falsely determines that a minimum has been found.

Figure 11–10
The Results Using
an Initial Parameter
Value of x_0 = 2.5

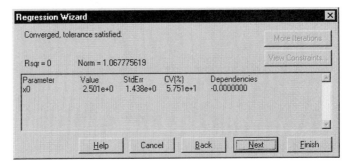

6. **Finding the absolute minimum**   Click Back, then click Options. Change the initial value of x_0 to 2.0.

7. Click OK to close the Options dialog box, then click Next to execute the regression. The initial parameter value is reasonably close to the optimum value, so the regression converges to the correct value $x_0 = 0.0$.

Figure 11–11
The Results Using
an Initial Parameter
Value of x_0 = 2.0

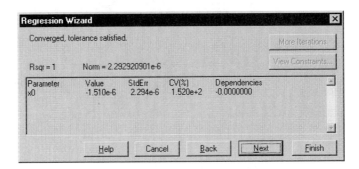

Summary   These last examples demonstrate how the curve fitter can find a local minimum and even a local maximum using poorly chosen initial parameter values.

Example 2: Weighted Regression

The data obtained from the lung washout of intravenously injected dissolved Xenon 133 is graphed in the Weighted Graph in the Weighted Regression section of the NONLIN.JNB notebook.

1. Open the Weighted worksheet and graph by double-clicking the graph page icon in the Weighted section of the NONLIN.JNB notebook.

   The data in the graph displays the compartmental behavior of Xenon in the body. Three behaviors are seen: the wash-in from the blood (rapid rise), the washout from the lung (rapid decrease), and the recirculation of Xenon shunted past the lung (slow decrease).

Figure 11–12
The Weighted Graph

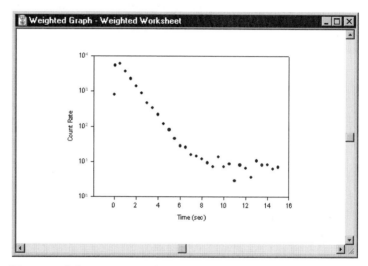

The sum of three exponentials (a triple exponential) is used as a compartmental model:

$$CountRate = a_1 e^{-1_1 t} + a_2 e^{-1_2 t} + a_3 e^{-1_3 t}$$

Least squares curve fitting assumes that the standard deviations of all data points are equal. However, the standard deviation for radioactive decay data increases with the count rate. Radioactive decay data is characterized by a Poisson random process, for which the mean and the variance are equal. Weighting must be used to account for the non-uniform variability in the data. These weights are the reciprocal of the variance of the data.

For a Poisson process, the variance equals the mean. You can use the inverse of the measurements as an estimate of the weights. The initial weighting variable only needs to be proportional to the inverse variance.

2. Double-click the Weighted Triple Exponential equation in the Weighted Regression section.

Figure 11–13
The Weighted Triple
Exponential Equation

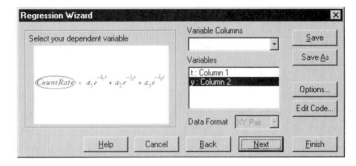

Click the Edit Code button, and examine the Variable value

w = 1/(col)2

This sets w to equal the reciprocal of the data in column 2. Click Cancel to close the dialog box.

3. Click the datapoints to select your variables. To use the w variable as the weighting variable, click Options, and select w as the Fit With Weight value.

Figure 11–14
Selecting a Weight Variable

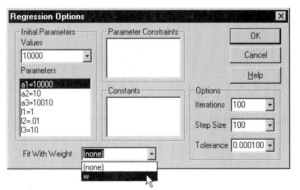

Click OK to close the dialog box.

4. Click Next to run the regression. The curve fitter finds a solution quickly. Click Finish to complete the regression.

5. What would be the result without weighting? Press F5, then click Next and click Options. Change the weighting to (none), then click OK.

6. Click Finish. The curve fitter goes through many more iterations. When it is completed, view the Weighted graph page.

The graph shows the nonlinear regression results with and without weighting. The weighted results fit the very small recirculation data (represented by the third exponential) quite well. However, when weighting is not used, the curve fitter ignored the relatively small values in the recirculation portion of the data, resulting in a poor fit.

Figure 11–15
Comparing the Function
Results of Weighted
and Unweighted
Nonlinear Regression Fits

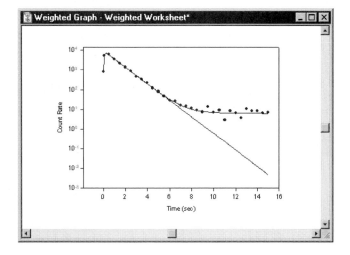

Example 3: Piecewise Continuous Function

The data obtained from the wash-in of a volatile liquid into a mixing chamber is modeled by three separate equations, representing three line segments joined at their endpoints:

$$f_1(t) = x_1(T_1 - t) + \frac{x_2(t - t_1)}{(T_1 - t_1)}$$

$$f_2(t) = x_2(T_2 - t) + \frac{x_3(t - T_1)}{(T_2 - T_1)}$$

$$f_3(t) = x_3(t_4 - t) + \frac{x_4(t - T_4)}{(t_4 - T_2)}$$

where:

$$f = \begin{cases} f_1(t) \text{ if } t_1 < t < T_1 \\ f_2(t) \text{ if } T_1 < t < T_2 \\ f_3(t) \text{ if } T_2 < t < t_4 \end{cases}$$

1. Open the Piecewise Continuous worksheet and graph by double-clicking the graph page icon in the Piecewise Continuous section of the NONLIN.JNB notebook. The data appears to be described by three lines, representing the three regions: before wash-in, during wash-in, and following wash-in.

2. View the notebook, and double-click the Piecewise Continuous Regression Equation. Click the datapoints to select the data, then click Next to run the regression. The model, with parameters x_1, x_2, x_3, x_4, T_1, and T_2, is fit to the data.

Figure 11–16
The Weighted Triple
Exponential Equation

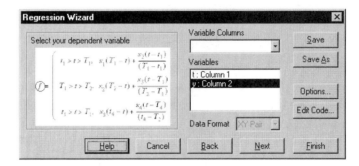

3. Click Finish. When the fit is complete, view the graph page. A continuous curve fits the three segments of the data.

Figure 11–17
The Data in the Piecewise
Continuous Graph Fitted
with the Equations
for Three Lines

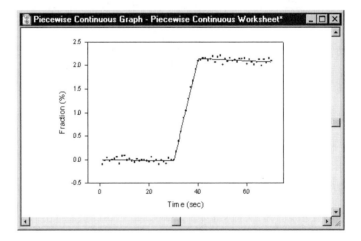

Example 4: Using Dependencies

This example demonstrates the use of dependencies to determine when the data has been "over-parameterized." Too many parameters result in dependencies very near 1.0. If a mathematical model contains too many parameters, a less complex model may be found that adequately describes the data.

Sums of exponentials are commonly used to characterize the dynamic behavior of compartmental models. In this example you model data generated from the sum of two exponentials with one, two, and three exponential models, and you examine the parameter dependencies in each case.

Dependencies Over a Restricted Range

The first fit is made to data over a restricted range, which does not reveal the true nature of the data.

1. Open the Dependencies worksheet and graph by double-clicking the graph page icon in the Dependencies section of the Regression Examples notebook. The data generated from the sum of two exponentials:

$$f(t) = 0.9e^{-t} + 0.1e^{-0.2t}$$

is graphed on a semi-logarithmic scale over the range 0 to 3.

Figure 11–18
The Dependencies
Graph Showing the Data
for the Sum of Two
Exponentials

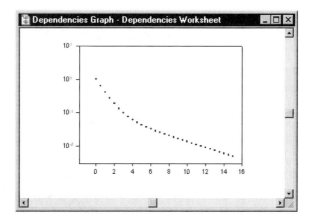

Although the data is slightly curved, the "break" associated with the two distinct exponentials is not obvious.

2. Right-click the curve and choose Fit Curve to open the Regression Wizard.

3. Select the Exponential Decay category and the Single, 2 Parameter exponential

decay equation, then click Next twice.

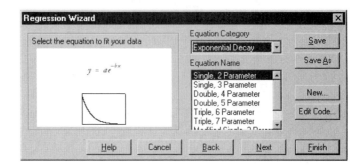

The results show that the dependencies are not near 1.0, indicating that the single exponential parameters, a_1 and b_1, are not dependent on one another.

Figure 11–20
The Results of Fitting
the Data
to a Single Exponential

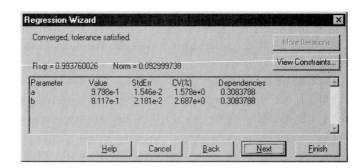

4. Click Back twice, and change the equation to the Double, 4 parameter exponential decay equation. Click Next twice.

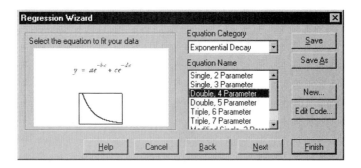

The results show that the parameter dependencies for the double exponential are acceptable, indicating that they are unlikely to be dependent, and that using a double exponential produces a better fit (the curve fitter in fact finds the exact parameter values used to generate the data, producing a perfect fit with an R^2 of 1).

Figure 11–22
The Results of Fitting the
Data to the
Sum of Two Exponentials

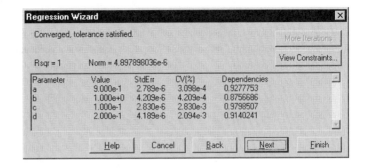

Dependencies Over an Extended Range

5. Click Back twice, and change the equation to a Triple, 6 Parameter exponential decay equation. Click Next twice.

Figure 11–23
Selecting the 6 Parameter
Triple Exponential Decay
Equation

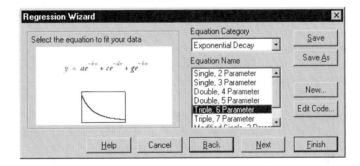

The results show that the parameter dependencies for a, b, c, and d are 1.00, suggesting that the three exponential model is too complex and that one exponential may be eliminated. Click Cancel when finished.

Figure 11–24
The Results of Fitting the
Data to the
Sum of Three Exponentials

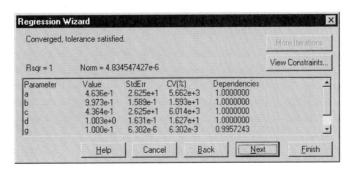

Example 5: Solving Nonlinear Equations

You can use the nonlinear regression to solve nonlinear equations. For example, given a y value in a nonlinear equation, you can use the nonlinear regression to solve for the x value by making the x value an unknown parameter.

Consider the problem of finding the LD_{50} of a dose response experiment. The LD_{50} is the function of the four parameter logistic equation:

$$f(x) = \frac{a_1}{1+e^{b(x-c)}} + d$$

where x is the dose and $f(x)$ is the response, then using nonlinear regression, you can find the value for x where:

$$50 = \frac{a_1}{1+e^{b(x-c)}} + d$$

1. Open the Solving Nonlinear Equations worksheet and graph file by double-clicking the graph page icon in the Solving Nonlinear Equations section of the NONLIN.JNB notebook. Note that the value for x at $y = 50$ appears to be approximately 150.

Figure 11–25
The Solving Nonlinear
Equations Graph, a Four
Parameter Logistic Curve

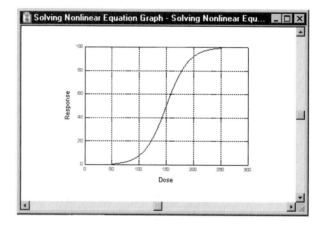

2. Double-click the Solving Nonlinear Equation and click the Edit Code button.

Figure 11–26
The Solving Nonlinear
Equations Statements Used
to Solve Four Parameter
Logistic Equation with Known
Parameters

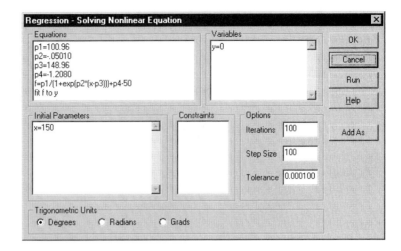

3. Examine the regression statements. Note that x is a parameter, y = 0, and the fit equation is modified:

$$f = p1/(1 + \exp(p2*(x−p3))) + p4 − 50$$

Since you are fitting f to y = 0, these statements effectively solve the original problem for *x* when *y* = 50. The values for parameters *a*, *b*, *c*, and *d* were obtained by fitting the four parameter logistic equation to a given set of dose response data.

4. Click Run to execute the regression. The parameter solution is the unknown *x*. For this example, *x* is approximately 149.5.

Figure 11–27
The Results the Solving
Nonlinear Equations Example

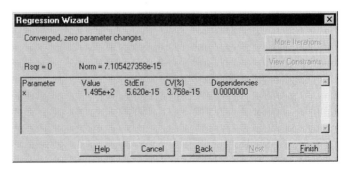

Example 6: Multiple Function Nonlinear Regression

You can use the Regression Wizard to fit more than one function at a time. This process involves combining your data into additional columns, then creating a third column which identifies the original data sets.

This example fits three separate equations to three data sets.

$$f_1(x) = \frac{T\left(\dfrac{x}{E_1}\right)^n}{1 + \left(\dfrac{x}{E_1}\right)^n} \ , \ f_2(x) = \frac{T\left(\dfrac{x}{E_2}\right)^n}{1 + \left(\dfrac{x}{E_2}\right)^n} \ , \ f_3(x) = \frac{T\left(\dfrac{x}{E_1}\right)^n}{1 + \left(\dfrac{x}{E_3}\right)^n}$$

1. Open the Multiple Function worksheet and graph by double-clicking the graph page icon in the Multiple Function section of the NONLIN.JNB notebook. The data points are for three dose responses.

    Columns 1 and 2 hold the combined data for the three curves. Column 3 is used to identify the three different data sets. A 0 corresponds to the first dataset, 1 to the second, and 2 to the third.

Figure 11–28
The Multiple Function
Graph with Three
Curves

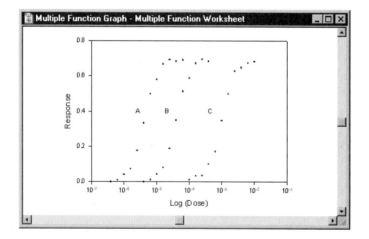

2. Double-click the Multiple Functions Equation. The Regression Wizard opens with the variables panel displayed. Click Edit Code.

Figure 11–29
The Multiple
Function Statements

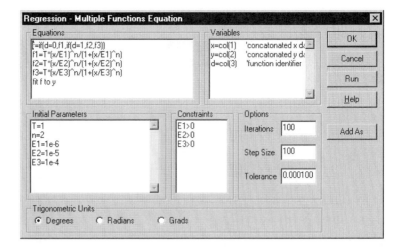

3. Examine the fit statements. The fit equation is an if statement which uses different equations depending on the value of d, which is the data set identifier variable. If $d = 0$, the data is fit to f1: if $d = 1$, the data is fit to f2; and if $d = 2$, the data is fit to f3.

   The functions share the T and n parameters, but have individual E parameters of E_1, E_2, and E_3.

4. Click Run to execute the regression. The fit proceeds slowly but fits each data set to the separate equation. Click Next to ensure that the Predicted function results are saved to the worksheet, then Next again and make sure no graph is being created. Click Finish to end the fit.

5. To graph the results, you need to create a plot of the predicted results. View the page and select the graph, then create a straight line plot of rows 1-12 of column

1 versus rows 1-12 of the predicted results column.

Figure 11–30
Creating a Plot of a
Restricted Data
Range

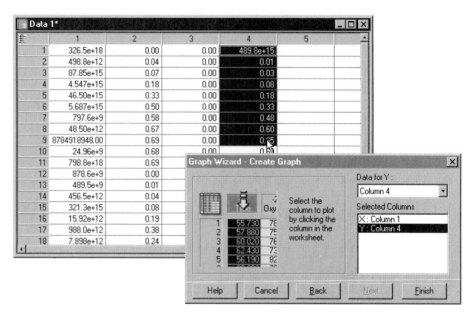

6. Create two more line plots of rows 13-23 and 24-34. The results plots should appear as three separate curves.

Figure 11–31
A SigmaPlot Graph
of the Predicted
Results of the
Multiple Function
Equation

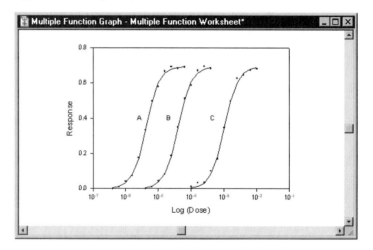

Example 7: Advanced Nonlinear Regression

Consider the function:

$$f = 1 - e^{\frac{-dx(b + cx)}{x + a}}$$

When fitted to the data in columns 1 and 2 in the Advanced Techniques worksheet, this equation presents several problems:

➤ Parameter identifiability

➤ Very large x values

➤ Very large y error value range

These problems are outlined and solved below.

If you want to view the regression functions for this equation, open the Advanced Techniques worksheet and graph in the Advanced Techniques section of the NONLIN.JNB notebook. Double-click the Advanced Techniques Equation to open the Regression Wizard. If you want to run the equation, use the graph of the transformed data.

Overparameterized Equations

The equation has four parameters, *a, b, c,* and *d.* The numerator in the exponential:

$$-dx(b + cx)$$

can have identical values for an infinite number of possible parameter combinations. For example, the parameter values:

$b = c = 1$ and $d = 2$

and the values:

$b = c = 2$ and $d = 1$

result in identical numerator terms.

The curve fitter cannot find a unique set of parameters. The parameters are not uniquely *identifiable,* as indicated by the large values for variance inflation factor (VIF), and dependency values near 1.0.

The solution to this problem is to multiply the *d* parameter with the other terms to create the equation:

$$f = 1 - e^{\left[\frac{-x(db + dcx)}{x + a}\right]}$$

then treat the *db* and *dc* terms as single parameters. This reduces the number of parameters to three.

Scaling Large Variable Values

The data used for the fit has enormous x values, around a value of 1×10^{24} (see column 1 in the worksheet above). These x values appear in the argument of an exponential which is limited to about ± 700, which is much smaller than 10^{24}. However, when the curve fitter tries to find the parameter values which are multiplied with x, it does not try to keep the argument value within ± 700. Instead, when the curve fitter varies the parameters, it overflows and underflows the argument range, and does not change the parameter values.

The solution to this problem is to scale the x variable and redefine some of the parameters. Multiply and divide each x value by 1×10^{24} to get:

$$f = 1 - e^{\left[\dfrac{-\dfrac{10^{-24}x}{10^{-24}}\left(db + \dfrac{dc\,10^{-24}x}{10^{-24}}\right)}{\dfrac{10^{-24}x}{10^{-24}} + a}\right]}$$

If you let $X = x(10^{-24})$, then the equation becomes:

$$f = 1 - e^{\left[\dfrac{-X(db + dc\,10^{24}X)}{X + 10^{-24}a}\right]}$$

If you let $CD = 10^{24}dc$ and $A = 10^{-24}a$, the resulting scaled equation is simplified to:

$$f = 1 - e^{\left[\dfrac{-X(db + CDX)}{X + A}\right]}$$

The exponent argument now does not cause underflows and overflows.

The graph of the transformed x data is displayed below the original data.

Figure 11–32
The Graph Page
for the Advanced
Techniques Example

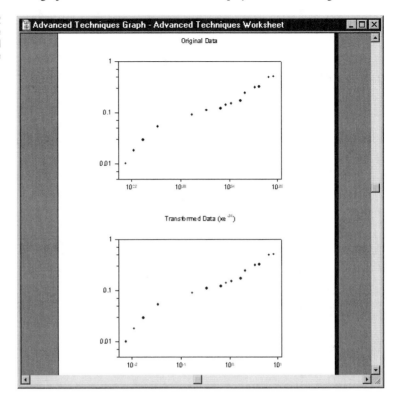

Small Independent Variable Values: Weighting for Non-Uniform Errors

The y values for the data range from very small values to very large values. However, for this problem, we know that the y values do not have the same errors—smaller y values have smaller errors.

The curve fitter fits the data by minimizing the sum of the squares of the residuals. Because the squares of the residuals extend over an even larger range than the data, small residual squared numbers are essentially ignored.

The solution to this non-uniform error problem is to use weighting, so that all residual squared terms are approximately the same size.

Fitting with a weighting variable of $1/y^2$ (the inverse of y squared), which is proportional to the inverse of the variance of the y data, produces a better fit for low y value data.

Figure 11–33
The Results of the
ADVANCED.FIT
with Weighting

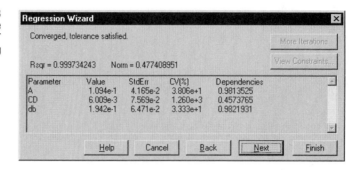

To see the results of the regression without weighting, open the Options dialog box and change the weighting to (none) before finishing.

Figure 11–34
The Graph Showing the
Results of Weighted
and Unweighted
Nonlinear Regressions

The dotted line indicates the
unweighted fit.

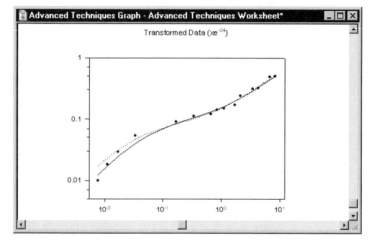

12

Automating Routine Tasks

SigmaPlot uses a VBA®-like macro language to access automation internally. However, whether you have never programmed, or are an expert programmer, you can take advantage of this technology by using the Macro Recorder.

This chapter describes how to use SigmaPlot's Macro Recorder and integrated development environment (IDE). It also contains descriptions of related features accessible in the Macro window, including the Sax Basic programming language, debugging tool, dialog box editor, and user-defined functions.

This chapter contains the following topics:

Creating Macros

Record a macro any time that you find yourself regularly typing the same keystrokes, choosing the same commands, or going through the same sequence of operations.

Before you Record   **Before you record the macro:**

1. Analyze the task you want to automate. If the macro has more than a few steps, write down an outline of the steps.

2. Rehearse the sequence to make sure you have included every single action.

3. Decide what to call the macro, where to assign it, and where to save it.

Recording a Macro   To learn how to place macros on the menu, see "Creating Macros as Menu Commands" on page 241.

To record a macro:

1. Choose Tools/Macro/Record New Macro.

    The REC appears in the status area of SigmaPlot's main window, indicating that the macro is recording your menu selections and keystrokes.

Figure 12–1
The Status Bar

2. Complete the activity you want to include in this macro.

    Note that the Macro Recorder does not record cursor movements.

3. Choose Tools/Menu/Stop Recording when you are finished recording the macro.

    The Macro Recorder stops recording and the Macro Options dialog box appears.

Figure 12–2
Macros Options
Dialog Box

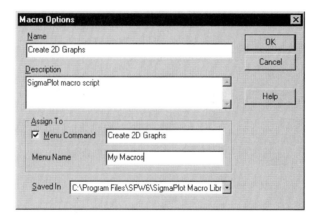

4. Type a name for the macro in the Name text box.

    Give the macro a descriptive name. You can use a combination of upper- and lowercase letters, numbers, and underscores. For example a macro that formats all of your graph legends to match a certain report might be called "Report1AddFormatToLegend".

5. Enter a more detailed description in the Description text box.

6. Click OK.

    After you have finished recording the macro, save it globally (for use in all of Sig-maPlot) or locally (for use in a particular notebook file).

When you return to the Notebook window, your macro appears in the Notebook.

Figure 12–3
Macros associated with notebooks appear in the notebook. Double-click a macro icon to open the Macros dialog box.

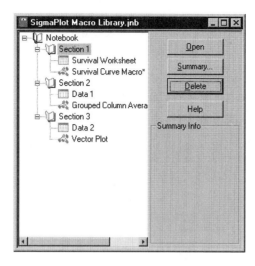

Creating Macros as Menu Commands

You can place your macro as a menu command on the main menu that you specify. For example, your new macro could appear on the main menu under the macro command "My Macros."

To create a new menu command:

1. Choose Tools/Macro/Macros.

    The Macros dialog box appears.

Figure 12–4
Macros Dialog Box

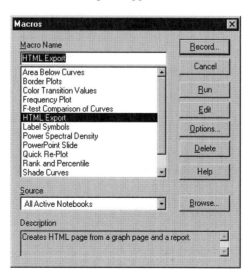

2. Select a macro from the Macro Name scroll-down list.

3. Click Options.

The Macro Options dialog box appears.

Figure 12–5
Macros Options Dialog Box

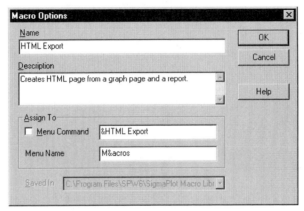

4. Select Menu Command.

5. Enter the name of the macro in the Menu Command field.

   If the Menu Command is cleared, the macro doesn't appear on a menu.

6. Enter the name of the menu under which you want the macro to appear in the Menu Name field.

7. Click OK.

   Your new macro appears under the menu command you've just created.

8. Enter the same menu command name in the Menu Name field of future macros if you want them to appear on your new macro command menu.

   By default, if the Menu Name field is left empty, the macro name appears on the Tools menu. You can also create your own menu by entering the menu name in the Menu Name field.

Running Your Macro

After you have recorded and saved a macro, it's ready to run.

To run a macro from the Macros dialog box:

1. Choose Tools/Macro/Macros.

The Macros dialog box appears with a list of available macros.

Figure 12–6
Macros Dialog Box

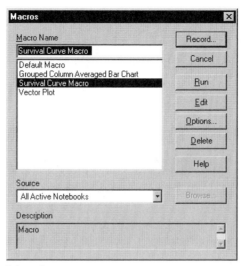

2. Select the macro to run.

3. Click Run.

To run a macro from a notebook:

1. From within a notebook, double-click the macro icon.

   The Macro dialog box appears with the corresponding macro selected.

2. Click Run.

   If the macro does not have any errors or run into difficulties with your data, it will run to completion.

Σ You can also run a macro from the Macro script window. This is useful for debugging the macro script.

Editing Macros

When you record a macro, SigmaPlot generates a series of program statements that are equivalent to the actions that you perform. These statements are in a form of SigmaPlot language that has custom extensions specifically for SigmaPlot automation and appear in the Macro Window. You can edit these statements to modify the actions of the macro. You can also add comments to describe code.

To edit a macro:

1. Choose Tool/Macro/Macros.

2. The Macros dialog box appears.

3. Select a macro from the Macro list.

4. Click Edit.

5. The Macro Window appears.

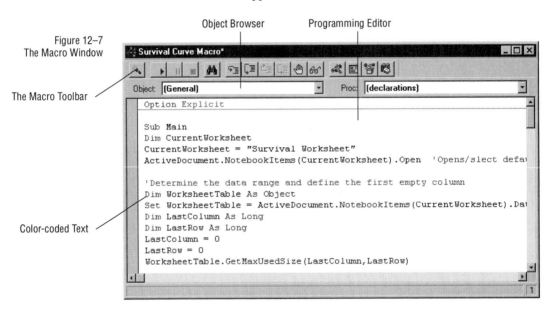

Figure 12–7
The Macro Window

Object Browser   Programming Editor

The Macro Toolbar

Color-coded Text

Using the Macro Window Toolbar The Macro Window toolbar appears at the top of the Macro Window. It contains buttons grouped by function.

Figure 12–8
The Macro Window Toolbar

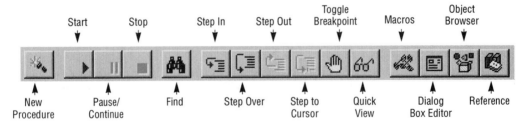

The following table describes the functions of the toolbar buttons in the Macro Window.

Toolbar button	Description
New Procedure	Opens the Add Procedure dialog box that lets you name the procedure and paste procedure code into your macro file.
Start	Runs the active macro and opens the Debug Window.
Pause/Continue	Pauses and restarts a running macro. This button also pauses and restarts recording of SigmaPlot commands while using the Macro Recorder.
Stop	Terminates recording of SigmaPlot commands in the Macro Recorder. Also, stops a running macro.
Find	Opens the Find dialog where you can define a search for text strings in the Macro Window.
Step In	Executes the current line. If the current line is a subroutine or function call, execution will stop on the first line of that subroutine or call.
Step Over	Executes to the next line. If the current line is a subroutine or a function call, execution of that subroutine or function call will complete.
Step Out	Steps execution out of the current subroutine or function call.
Step to Cursor	Steps execution out to the current subroutine or function call.
Toggle Breakpoint	Toggles the breakpoint on the current line. The breakpoint stops program execution.
Quick View	Shows the value of the expression under the cursor in the Immediate Window.
Toggle Comments	Hides and shows the comments in your macro code.
Macros	Opens the Macros dialog box.
Dialog Box Editor	Opens the Dialog Box Editor.
Object Browser	Opens the Object Browser.
Reference	Opens the Reference dialog box which contains a list of all programs that are extensions of the SigmaPlot Basic language.

Color-Coded Display   The color-coding of text in the Macro Window indicates what type of code you are
viewing.

The following table describes the default text colors used in the script text:

Text Color	Description
Blue	Identifies reserved words in Visual Basic (for example, Sub End Sub, and Dim).
Magenta	Identifies SigmaPlot macro commands and functions.
Green	Identifies comments in your macro code. Separates program documentation from the code as you read through your macros.

Object and
Procedure Lists   The Object and Procedure lists show SigmaPlot objects and procedures for the
current macro. These lists are useful when your macros become longer and more
complex.

➤   The object identified as "(General)" groups all of the procedures that are not
part of any specific object.

➤   The Procedure list shows all of the procedures for the currently selected object.

Setting Macro
Window Options   You can set appearance options for the Macro window in the Macros tab of the
Options dialog box.

To set the options of the Macro Window:

1.   With a macro window open, Choose Tools/Options.

The Options dialog box appears.

Figure 12–9
The Options Dialog Box
Macro Tab

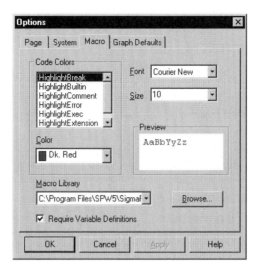

2. Click the Macros tab.

3. Set text colors for different types of macro code and Debug Window output.

4. Change font characteristics.

5. Set the location for the macro library.

Parts of the Macro Programming Language

The following topics list the parts of the macro programming language:

➤ *Statements* are instructions to SigmaPlot to perform an action(s). Statements can consist of *keywords*, *operators*, *variables*, and *procedure* calls.

Keywords are terms that have special meaning in SigmaPlot. For example, the Sub and End Sub keywords mark the beginning and end of a macro. By default, keywords appears as blue text on color monitors. To find out more about a specific keyword in a macro, select the keyword and press F1. When you do this, a topic in the SigmaPlot on-line reference appears and presents information about the term.

➤ You can add optional comments to describe a macro command or function, and how it interacts in the script. When the macro is running, comment lines are ignored. Indicate a comment by beginning a line with an apostrophe. Comments always must end the line they're on. The next program line must go on a new line. By default, comment lines appear as green text.

Scrolling and moving the insertion point

When you use the scroll bars the insertion point does not change. To edit the macro code that you are viewing in the macro window, you must move the insertion point manually.

To edit macro code manually:

1. In the Macro window, click where you want to edit.

2. You can also use arrows and key combinations to move the insertion point; when you do this the window scrolls automatically.

Editing macro code

You can edit macro code in the same way you edit text in most word-processing and text editing programs. You add select and delete text, type over code, or insert text by moving the insertion point and then typing in new text. As with other programming languages, you can also add comments to code.

To edit macro code:

➤   Open the macro code window and select the text to edit.

Adding Comments to Code

Adding comments to code is an excellent way to identify the purpose of the various parts of a macro and to map locations as you edit a complex macro. Comments can be inserted that fully document how to use and how to understand the macro code.

Deleting Unnecessary Code

The Macro Recorder creates code corresponding exactly to the actions that you make in SigmaPlot while the recorder was turned on. You may need to edit out unwanted steps.

Moving and Copying Code   You can cut, copy, and paste selected text.

Finding and Replacing Code   When you need to find and change text in a macro that you have written, use the Find commands. For example, if you change the name of a file that is referenced in your macro, you need to change every instance of the file name in your macro. Use Find to locate the instances of the filename in the macro and replace using cut and paste edit commands.

Adding Existing Macros to a Macro

If you have another macro that already does what you want, you can just paste it into your new macro. Copy and paste the macro into your new macro, test it in the new code and run it.

What Macro Recorder records

The Macro Recorder does not record the following types of activity:

➤   Cursor movement

If you want to include this type of activity in your macro, you can use the IDE features.

About user-defined functions

A user-defined function is a combination of math expressions and Basic code. The function always requires input data values and always returns a value. You supply the function with a value; it performs calculations on the values and returns a new value as the answer. Functions can work with text, dates, and codes, not just numbers.

A user-defined function is similar to a macro but there are differences. Some of the differences are listed in the following table.

Recorded macros	User-defined functions
Performs a SigmaPlot action, such as creating a new chart. Macros change the state of the program.	Returns a value; cannot perform actions. Functions return answers based on input values.
Can be recorded.	Must be created in Macro code.
Are enclosed in the Sub and End Sub keywords.	Are enclosed in the keywords Function and End Function.

For more information
: The on-line help has an extensive section on user-defined functions. From anywhere in the Macro window, press F1, or choose Help Topics from the Help menu.

Creating user-defined functions
: A user-defined function is like any of the built-in SigmaPlot function. Because you create the user-defined function, however, you have control over exactly what it does. A single user-defined function can replace database and spreadsheet data manipulation with a single program that you call from inside SigmaPlot. It is a lot easier to remember a single program than it is to remember several spreadsheet macros.

For a full explanation of User Defined Functions, see the Automation on-line reference Help file.

Using the Dialog Box Editor

The Dialog Box Editor lets you design and customize your own dialog boxes. When you are designing and creating SigmaPlot macros, you can automatically create the necessary dialog box code and dialog monitor function code with the Dialog Box

Editor. Like the other automated coding features in SigmaPlot, the code may require further customization.

To create a custom dialog box:

1. In the Macro Window, place the insertion point where you want to put the code for the dialog box.

2. From the Macro Window toolbar click the Dialog Box Editor button. A blank dialog grid appears.

3. Now you can select a tool, such as a button or check boxes, from the Toolbox. The cursor changes to a cross when you move it over the grid.

4. To place a tool on the dialog box, click a position on the grid. A default tool will be added to the dialog grid.

5. Resize the dialog box by dragging the handles on the sides and the corners.

6. Right-click any of the controls that you have placed on the dialog surface (after selecting the control) and enter a name for the control.

7. Right-click the dialog form (with no control selected) and enter a name for the dialog monitor function in the DialogFunc field.

8. To finish, click OK. The code for the dialog box with controls will be written to the Macro Window.

9. Finally, and in most cases, you must edit the code for dialog box monitor function to define the specific behavior of the elements in your dialog box.

For more information, see the Automation on-line help reference.

Using the Object Browser

The Object Browser displays all SigmaPlot object classes. The methods and properties associated with each SigmaPlot macro object class are listed. A short description of each object appears in the dialog box as you select them from the list. By clicking F1, you can access extensive Help that includes example code for the individual properties and methods. The Paste feature lets you insert generic code based on your selection into a macro.

The Object Browser will be familiar and useful if you are comfortable with object oriented programming. If you are not, consult one of the excellent introductory texts on Visual Basic.

For full details on using the Object Browser, press F1 from anywhere in the Macro window.

Using the Add Procedure Dialog Box

Organizing your code in procedures makes it easier to manage and reuse. SigmaPlot macros, like Visual Basic programs, must have at least one procedure (the main subroutine) and often they have several. The main procedure may contain only a few statements, aside from calling subroutines that do the work.

SigmaPlot provides a dialog box that generates procedure code for your macros.

Using the Add Procedure Dialog Box

By using the Add Procedure dialog box, you can define a sub, function, or property using the Name, Type, and Scope boxes. Clicking OK pastes the code for a new procedure into your macro at the insertion point.

For full details on using the Add Procedure, press F1 from anywhere in the Macro window.

Using the Debug Window

The Debug Window contains a group of features that are helpful when you are trying to locate and resolve errors in your macro code. The debugging tools in SigmaPlot will be familiar if you have used one of the modern visual programming languages or Microsoft Visual Basic for Applications. Essentially, the Debug Window gives you incremental control over the execution of your program so that you can sleuth errors in your programs. The Debug Window also gives you a precise way to determine the contents of your variables. Again, a series of buttons is used to select the operation mode of the Debug Window.

Debug Toolbar Buttons

The debugging features of the Debug Window are controlled by buttons on the Macro Window toolbar. To review:

➤ The four Step buttons provide methods for controlling the execution of commands. They offer various ways of responding to subroutines and functions.

➤ The Breakpoint button lets you set a point and execute the program until it reaches that point.

➤ The Quick View button displays the value of the expression in the immediate window.

The inclusion of these features for controlling program execution are a standard but powerful combination of tools for writing and editing macros.

Debug Window Tabs

The output from the Debug Window is organized in four tabs that allow you to type in statements, observe program execution responses, and iteratively modify your code using this feedback. If you have never used a debugging tool and are new to programming, it would be a good idea to supplement the following description with further study.

Immediate Tab

The Immediate Tab lets you evaluate an expression, assign a specific value to a variable or call a subroutine and evaluate the results. Trace mode prints the code in the tab when the macro is running.

➤ Type "?expr" and press Enter to show the value of "expr".

➤ Type "var = expr" and press Enter to change the value of "var".

➤ Type "set var = expr" and press Enter to change the reference of "var".

➤ Type "subname args" and press Enter to call a subroutine or built-in expression "subname" with arguments "args".

➤ Type "trace" and press Enter to toggle trace mode. Trace mode prints each statement in the Immediate Tab when a macro is running.

Watch Tab

The Watch Tab lists variables, functions, and expressions that are calculated during execution of the program.

➤ Each time program execution pauses, the value of each line in the window is updated.

➤ The expression to the left of the "->" may be edited.

➤ Pressing Enter updates all the values immediately.

➤ Pressing Ctrl+Y deletes the line.

Stack Tab

The output from the Stack Tab lists the program lines that called the current statement. This is a macro command audit and is helpful to determine the order of statements in you program.

➤ The first line is the current statement. The second line is the one that called the first, and so on.

➤ Clicking a line brings that macro into a sheet and highlights the line in the edit window.

13

SigmaPlot Automation Reference

OLE Automation is a technology that lets other applications, development tools, and macro languages use a program. SigmaPlot Automation allows you to integrate SigmaPlot with the applications you have developed. It also provides an effective tool to customize or automate frequent tasks you want to perform.

Automation uses objects to manipulate a program. Objects are the fundamental building block of macros; nearly all macro programs involve modifying objects. Every item in SigmaPlot-graphs, worksheets, axes, tick marks, reports, notebooks, etc.-can be represented by an object.

SigmaPlot uses a VBAÆ-like macro language to access automation internally. For more information on recording and editing SigmaPlot macros, see About Macros.

This chapter contains the following topics:

➤ "Opening SigmaPlot from Microsoft Word or Excel" on page 253
➤ "SigmaPlot Objects and Collections" on page 254
➤ "SigmaPlot Properties" on page 264
➤ "SigmaPlot Methods" on page 273

Opening SigmaPlot from Microsoft Word or Excel

To open SigmaPlot from Microsoft Word or Excel, you must first create a macro from within either application.

To create the macro:

1. In Excel or Word, choose Tools/Macro/Visual Basic Editor.

   Visual Basic appears.

2. Choose Insert/Module.

3. Type:

```
Sub SigmaPlot()
'
```

```
'  SigmaPlot Macro
'
'
    Dim SPApp as Object
    Set SPApp = CreateObject("SigmaPlot.Application.1")
    SPApp.Visible = True

    SPApp.Application.Notebooks.Add
End Sub
```

4. Choose Run/Run Sub/User Form to run the macro.

SigmaPlot appears with an empty worksheet and notebook window.

To open SigmaPlot from Word or Excel in the future:

1. Choose Tools/Macro/Macros to open the Macros dialog box.

2. Select SigmaPlot.

3. Click Run.

SigmaPlot Objects and Collections

An *object* represents any type of identifiable item in SigmaPlot. Graphs, axes, notebooks, worksheets, and worksheet columns are all objects.

A *collection* is an object that contains several other objects, usually of the same type; for example, all the items in a notebook are contained in a single collection object. Collections can have methods and properties that affect the all objects in the collection.

Properties and *methods* are used to modify objects and collections of objects. To specify the properties and methods for an object that is part of a collection, you need to return that individual object from the collection first.

For more information, refer to SigmaPlot Automation Help from the SigmaPlot Help menu.

Application Object   An Application object represents the SigmaPlot application, within which all other objects are found. (Most other objects must exist inside higher-level objects. You access objects by applying properties and methods on these higher-level objects.) It is a "user-creatable" object, that is, outside programs can run SigmaPlot and access its properties directly, and will be registered in registry as SPW32.Application.

The Application object properties and methods return or manipulate attributes of the SigmaPlot application and main window, and access the list of notebooks and from there all other objects.

To use the Application Object

Use Application properties to return attributes of the SigmaPlot application. Note that when using the SigmaPlot macro window, all Application methods and properties are global, that is, you do not need to specify the Application object.

Notebooks Collection Object

The Notebooks collection represents the list of open notebooks in SigmaPlot. Use this collection to create new documents and open existing documents, as well as to specify and return individual notebooks as objects.

To use the NotebooksCollection

A Notebooks collection is returned using the Application object Notebooks property. Use the Add method to add a new notebook to the collection. You can return a specific Notebook object using either the Item property or the collection index

Notebook Object

Represents a SigmaPlot notebook file (including template and equation library files). Notebook properties and methods are used to set individual notebook file attributes and specify individual notebook items, e.g., worksheets, graph pages, reports, etc. Also used to return a collection of notebook items as objects.

To use the Notebook Object

Notebook objects are returned using the Notebooks or ActiveDocument Application object properties. Access individual notebook items using the NotebookItems property, which returns the NotebookItems collection.

NotebookItems Collection Object

This collection represents all the items in a notebook, and is used to create new items and open existing items. Also used to specify and return the different notebook items as objects. Worksheets, pages, equations, reports, macros, and section and notebook folders are all notebook items and can be returned as objects.

To use the NotebookItems Collection

The NotebookItems collection is returned using the NotebookItems property of a Notebook object. You can return individual notebook item objects using either the Item method or collection index, and add new notebook item objects, such as worksheets and graph pages, using the Add method.

NativeWorksheetItem Object

This object represents the SigmaPlot data worksheet. Use this object to perform worksheet edit operations, and to access the data using the DataTable property.

To use the NativeWorksheetItem Object

The NativeWorksheetItem object has the standard notebook item properties and methods. A NativeWorksheetItem is returned from the NotebookItems collection using the Item property or collection index, and created using the NotebookItems Add method. The NativeWorksheetItem object has an ItemType property and NotebookItems.Add method value of 1.

Excelltem Object   Use this object to manipulate in-place activated Excel worksheets. In general, most NativeWorksheetItem properties and methods also apply to Excel worksheets.

To use the Excelltem Object

The Excelltem object also has the standard notebook item properties and methods. An Excelltem is returned from the NotebookItems collection using the Item property or collection index, and created using the NotebookItems Add method. The Excelltem object has an ItemType property and NotebookItems.Add method value of 8.

Σ   Note that you can only create an Excelltem if you have Excel for Office 95 or 97 installed.

DataTable Object   Represents a table of data as used by a worksheet or graph page. This object's properties and methods can be used to access the data in a worksheet or page, and also return the NamedRanges (row and column titles) collection object.

To use the DataTable Object

The DataTable objects is returned from NativeWorksheetItem, Excelltem, and GraphItem objects using the DataTable method, and in turn accesses data using the GetData and PutData methods and the Cell property.

NamedDataRanges
Collection Object   The NamedDataRanges collection contains all ranges in the DataTable object that have been assigned a name. Column and row titles are name ranges.

To use the NamedDataRanges Collection

The NamedDataRanges collection is mainly used to retrieve existing ranges and add new ranges to DataTable objects within NativeWorksheetItem and GraphItem objects.

NamedDataRange
Object   Represents named data range objects (e.g. column and row titles) in the worksheet and page data tables.

To use the NamedDataRange Object

The NamedDataRange object is returned from the NamedRanges collection using an index or the Item property. The NamedRange object properties are mainly the range name, dimensions and other similar attributes.

GraphItem Object   The GraphItem object represents a SigmaPlot graph page. GraphItem properties can be used to return a collection of the graphs on the page using the GraphObjects property. It can also be used to create graphs using the CreateWizardGraph method.

To use the GraphItem Object

The GraphItem object has the standard notebook item properties and methods. A GraphItem is returned from the NotebookItems collection using the Item property or collection index, and created using the NotebookItems Add method. The GraphItem object has an ItemType property and NotebookItems.Add method value of 2.

Pages Collection/ Page Object

The Page object represents a SigmaPlot graph page. Graph pages can be different sizes and colors, and a page object can be used to return the collections of objects on that page. Pages have an ObjectType value of 1 or GPT_PAGE.

To use the Page Object

A graph page is returned from a GraphItem object using the GraphPages property. Note that since there is currently only one page per graph item, you can always use .GraphPages(0) to return a page. Use the ChildObjects property to return the GraphObjects collection, or the Graphs property. Note that if a graph is part of a Group object, it can only be returned using the Graph property.

Many Page attributes and attribute values can only be returned or set using the GetAttribute and SetAttribute methods. Use the Page Attribute constants to specify these attributes.

Page GraphObjects Collection

The Page GraphObjects Collection represents a collection of the child objects returned from a Page object.

To use the Page GraphObjects Collection

A Page GraphObjects collection is returned from a Page object using the ChildObjects property. You can also return Page GraphObjects collections composed only of Graph objects using the Graphs property.

Graph Object

The Graph object represents a SigmaPlot graph. A graph is used to access the parts of a graph, e.g., plots, axes, etc., as well as to change graph attributes such as title, size, and position. Graph objects have an ObjectType value of 2 or GPT_GRAPH.

To use the Graph Object

A Graph object is returned from a Page GraphObjects collection. Note that you can create a GraphObjects collection composed only of the graphs using the Page object Graphs property. The ChildObjects or Plots properties are used to return the graph object's Plots collection. Other properties, such as Axes and AutoLegend, are used to return other graph objects.

Many Graph attributes and attribute values can only be returned or set using the GetAttribute and SetAttribute methods. Use the Graph Attribute constants to specify these attributes.

Plot GraphObjects Collection

Represents a collection of the child objects returned from a Graph object.

To use the Plot GraphObjects Collection

A Plot GraphObjects collection is returned from a Graph object using either the ChildObjects or Plots property. Use the Plots collection to return specific Plot objects.

Plot Object The Plot object represents a data plot and all its attributes and child objects. Plots have an ObjectType value of 3 or GPT_PLOT.

To use the Plot Object

Plot properties and methods are mainly used to return the Plot child objects. Return specific Plot child objects using different Plot properties:

➤ Line returns the Line plot object
➤ Symbols returns the Symbol plot object
➤ Fill returns the Solid plot object (e.g. bars)
➤ DropLines returns a collection of the drop line Line objects
➤ Functions returns collection of the Function objects (e.g. regression and reference lines)

Use .ChildObjects(0) to return the Tuple GraphObjects collection. Tuples represent the individual plot curves.

Many Plot attributes and attribute values can only be returned or set using the GetAttribute and SetAttribute methods. Use the Plot Attribute constants to specify these attributes.

Axes GraphObjects Collection The Axes collection corresponds to all the sets of axes available for a graph.

To use the Axes Collection

Use the index to return specific x, y or z axes from an Axes collection.

Axis Object The Axis objects represents a SigmaPlot axis. Axes have an ObjectType value of 4 or GPT_AXIS.

To use the Axis Object

An axis has several Line and Text objects associated with it, including the axis line itself, grid lines, tick marks and labels, and the axis title. Use the LineAttributes property to return the collection of axis lines, and the AxisTitles and TickLabelAttributes property to return collection of axis text objects.

Most other Axis attributes, such as range, scale, breaks, etc., can only be returned or set using the GetAttribute and SetAttribute methods. Use the Axis Attribute constants to specify these attributes.

Text Object

All characters and labels found on a SigmaPlot change correspond to a text object and can be modified using text object properties and methods. Axes have an ObjectType value of 5 or GPT_TEXT.

To use the Text Object

The Text object for most graph objects can be returned using the NameObject property.

Text objects are also found below both AutoLegend and Axis objects. Use the TickLabelAttributes property to access the tick label Text object. Use the AxisTitles property to access the axis titles.

To access the text objects within a Page or an AutoLegend, use the ChildObjects property.

Use the Name property to change the string used for the text. Most other Text attributes and attribute values can only be returned or set using the GetAttribute and SetAttribute methods. Use the Text Attribute constants to specify these attributes.

Line Object

Objects that correspond to drawn lines. These lines include all lines used for axes and plots, regression and reference lines, drop lines, and manually drawn lines. Lines have an ObjectType value of 6 or GPT_LINE.

To use the Lines Object

Lines are returned from a number of different objects:

Drawn lines on a page are returned using the ChildObjects property. Lines are creating using the Page object Add method with an object value of 6 or GPT_LINE.

Plot lines are returned using the Line property. Collections of Lines can also be returned using the DropLines property.

To return the collection of axis lines, use the Axis object LineAttributes property. These include the axis lines, tick marks, grid lines, and axis break lines.

The collection of function lines is returned using the Plot object Functions property. Functions include linear regression and reference lines.

Many Line attributes and attribute values can only be returned or set using the GetAttribute and SetAttribute methods. Use the Line Attribute constants to specify these attributes.

Symbol Object

The symbol object controls the symbols used in a plot. Symbols have an ObjectType value of & or GPT_SYMBOL.

To use the Symbols Object

The Symbols object is returned from a Plot object using the Symbols property. The Symbols properties and methods are used to return or modify the individual symbols within the parent plot.

Many Symbol attributes and attribute values can only be returned or set using the GetAttribute and SetAttribute methods. Use the Symbol Attribute constants to specify these attributes.

Solid Object

A solid object can represent many different graph and page objects. All drawn shapes-ellipses and rectangles, as well as graph bars and boxes, pie slices, meshes, and any other "filled" object. Solids have an ObjectType value of 8 or GPT_SOLID.

To use the Solid Object

Solid objects can be returned from a number of other groups or collections using different properties:

To access the solid object(s) for a Plot, use the Fill property.

To access the solid object(s) within a Page or an AutoLegend, use the ChildObjects property.

Many Solid attributes and attribute values can only be returned or set using the GetAttribute and SetAttribute methods. Use the Solid Attribute constants to specify these attributes.

Tuple GraphObjects Collection

Represents the collection of tuples for a Plot object. A tuple is an individual curve or series representing a plotted column or column set. For example, an XY pair plotted as a scatter plot, a single column plotted as a bar series, or a column plotted as a mean are all tuples.

To use the Tuples GraphObjects Collection

The Tuples collection is returned from a Plot using the ChildObjects property. Use the Tuple collection return specific tuples or add new tuples.

Tuple Object

A tuple is an object that represents a plotted column or column pair, displayed as a curve, datapoint, or bar series. A plot always consists of a collection of one or more tuples. Tuples have an ObjectType value of 9 or GPT_TUPLE.

To use the Tuple Object

Tuples are returned from the Tuples GraphObjects collection. Most generic GraphObject properties are not retained by tuples; instead, use the SetAttribute and GetAttribute methods to return or change the tuple properties. Use the Tuple Attribute constants to specify these attributes.

Function GraphObjects Collection

The Functions collection consists of all regression, confidence, prediction, and reference lines for a plot.

To use the Functions Collection

The functions collection is basically used to return individual function objects. Return the Functions Collection from a Plot object using the Functions property.

Function Object

A Function object represents one of the various function lines of a Plot object. In addition to Line object properties, functions also have properties and attributes specific to regression and reference lines. Functions have an ObjectType value of 10 or GPT_FUNCTION.

To use the Function Object

The Function object is returned from the Functions collection as follows:

Index	Constant	Function
1	SLA_FUNC_REGR	Regression Line
2	SLA_FUNC_CONF1	Upper Confidence Intervals
3	SLA_FUNC_CONF2	Lower Confidence Interval
4	SLA_FUNC_PRED1	Upper Prediction Interval
5	SLA_FUNC_PRED2	Lower Prediction Interval
6	SLA_FUNC_QC1	1st Reference Line (Upper Specification)
7	SLA_FUNC_QC2	2nd Reference Line (Upper Control Line)
8	SLA_FUNC_QC3	3rd Reference Line (Mean)
9	SLA_FUNC_QC4	4th Reference Line (Lower Control Line)
10	SLA_FUNC_QC5	5th Reference Line (Lower Specification)

Most Function attributes and attribute values can only be returned or set using the GetAttribute and SetAttribute methods. Use the Function Attribute constants to specify these attributes. You can return the label for reference lines using the NameObject property, but only if the label is turned on first.

Note that Function lines are turned on or off with Plot object attributes.

DropLines Collection

The DropLines object is a special collection of lines that represent the drop lines for a plot. The DropLines object is returned using the Plot object DropLines property. There are three different sets of drop lines that can be retrieved from the DropLines collection:

DropLine property indexes:

1. xy plane (SLA_FLAG_DROPZ , 3D graphs only)

2. Y axis/x direction or yz plane (SLA_FLAG_DROPX)

3. X axis/y direction or zx plane (SLA_FLAG_DROPY)

Note that drop lines are turned on and off using the Plot object SetAttribute method, using the SLA_PLOTOPTIONS property coupled with the SLA_FLAG_DROPX, SLA_FLAG_DROPY, or SLA_FLAG_DROPZ value, and using the FLAG_SET_BIT to turn on drop lines, or the FLAG_CLEAR_BIT to turn off drop lines. Other drop line properties are set using Line object attributes.

Group Object

A group is any grouped collection of objects, generally created with the Format menu Group command. Grouped objects can be treated as a single object. AutoLegends are a special class of Group object. Groups have an ObjectType value of 12 or GPT_BAG.

Many Group attributes and attribute values can only be returned or set using the GetAttribute and SetAttribute methods. Use the Group (Bag) Attribute constants to specify these attributes.

AutoLegend Object

The AutoLegend object is really a specific Group object consisting of a solid object and text objects. The AutoLegend object is returned from a graph using the AutoLegend property.

To use the AutoLegend Object

Manipulate the AutoLegend object as you would a Group object. Use the ChildObjects property to return the members of the AutoLegend. .ChildObjects(0) always returns the Solid rectangle object used as the AutoLegend border/background, and indexes >0 to return the individual Text objects used for the legend keys.

Legends can also be manipulated with many Text attributes using the GetAttribute and SetAttribute methods.

GraphObject Object

The GraphObject object corresponds to non-SigmaPlot objects residing on a graph page, such as pasted bitmaps or metafiles, or embedded or linked OLE objects.

FitItem Object

The FitIItem object corresponds to the a SigmaPlot equation and all the equation code parameters and settings. FitItems are used not only for regressions, but for other nonlinear curve fitting applications, and function plotting and solving. The results of a FitItem are accessed from a FitResults object.

To use the FitItem Object

The FitItem object has the standard notebook item properties and methods. A FitItem is returned from the NotebookItems collection using the Item property or collection index, and created using the NotebookItems Add method. The FittItem object has an ItemType property and NotebookItems.Add method value of 6.

The complete list of FitItem properties and methods also can be found in FitItem and FitResults Properties and Methods.

FitResults Object
The FitResults object is used to return the different values computed by the nonlinear regression. These statistics and other results are specifically useful for computing additional statistics that can be derived from these results.

The complete list of FitResult properties can also be found in FitItem and FitResults Properties and Methods.

TransformItem Object
Represents either an open transform or opened transform file as an object. You can load transforms, specify the transform code, and replace variables before executing a transform.

To use the TransformItem Object

The TransformItem object has the standard notebook item properties and methods; however transforms cannot be currently saved as notebook objects, only created and opened. If you want to save a transform to a .xfm file, use the Name property to specify a file name and path before using the Save method.

When using a TransfomItem object, you must first declare a variable as an object and then define it as a newly added transform item. Create a new TransformItem collection using the NotebookItems Add method, using a value of 9. After defining the transform object, open it using the Open method.

Specify the transform code using the Text property. To change the value of a transform variable, use the AddVariableExpression method. Run transforms using the Execute method.

Note: After executing a transform, it is a good idea to close it, as you are limited to four concurrent transforms that can be open simultaneously.

ReportItem Object
A ReportItem represents the RTF (Rich Text Format) documents used for text and regression reports in SigmaPlot. You can use the ReportItem properties to add and remove block of text from a report.

To use the ReportItem Object

The ReportItem object has the standard notebook item properties and methods. A ReportItem is returned from the NotebookItems collection using the Item property or collection index and created using the NotebookItems Add method. SigmaPlot reports have an ItemType property and NotebookItems.Add method value of value of 5, and SigmaStat reports have a value of 4.

MacroItem Object
Represents a SigmaPlot macro. You can use this command to edit and run macros from within macros, or to run macros from outside applications.

To use the MacroItem Object

The MacroItem object has the standard notebook item properties and methods. A MacroItem is returned from the NotebookItems collection using the Item property or collection index, and created using the NotebookItems Add method. The MacroItem object has an ItemType property and NotebookItems.Add method value of 0.

NotebookItem Object Represents the notebook item in the notebook window. You can use this object to rename the notebook item. The notebook item can always be reference with NotebookItems(0).

To use the NotebookItem Object

The NotebookItem object has most of the standard notebook item properties and methods, and created using the NotebookItems Add method. The NotebookItem object has an ItemType property and NotebookItems.Add method value of 7.

SectionItem Object Represents the section folders within a SigmaPlot notebook.

To use the SectionItem Object

The SectionItem object most of the standard notebook item properties and methods. A SectionItem is returned from the NotebookItems collection using the Item property or collection index, and created using the NotebookItems Add method. The SectionItem object has an ItemType property and NotebookItems.Add method value of 3.

SigmaPlot Properties

A *property* is a setting or other attribute of an objectóthink of a property as an "adjective." For example, properties of a graph include the size, location, type and style of plot, and the data that is plotted. To change the settings of an object, you change the properties settings. Properties are also used to access the objects that are below the current object in the hierarchy.

To change a property setting, type the object reference followed with a period, then type the property name, an equal sign (=), and the property value.

For more information, refer to SigmaPlot Automation Help from the SigmaPlot Help menu.

Application Property Used without an object qualifier, this property returns an Application object that represents the SigmaPlot application. Used with an object qualifier, this property returns an Application object that represents the creator of the specified object (you can use this property with an Automation object to return that object's application).

Σ Use the CreateObject and GetObject functions give you access to an Automation object.

Author Property   A standard property of notebook files and all NotebookItems objects. Returns or sets the Author field in the Summary Information for all notebook items, or the Author field under the Summary tab of the Windows 95/98 file Properties dialog box.

Autolegend Property   Returns the Autolegend Group object for the specified Graph object. Autolegends have all standard group properties. The first ChildObject of a legends is always a solid; the successive objects are text objects with legend symbols.

Axis Property   The Axes property is used to return the collection of Axis objects for the specified graph object. Individual axis objects have a number of line and text objects that are returned with Axis object properties.

Axistitles Property   The AxisTitle property is used to return the collection of axis title Text objects for the specified Axis. Use the following index values to return the different titles. Note the specific title returned depends on the current axis dimension/direction selected.

0	Bottom/Left axis title
1	Right/Top axis title
2	Sub axis title (not currently shown)
3	Sub axis title (not currently shown)

Cell Property   Returns or sets the value of a cell with the specified column and row coordinates for the current DataTable object.

ChildObjects Property   Used by all page objects that contain different sub-objects to return the collection of those objects. The ChildObjects property returns different type of objects depending on the object type:

Object Returns	ChildObjects
Page	Page GraphObjects
Graph	Plots
Plot	Tuples
Tuples	Tuple
Group (including Autolegends)	all group objects

Color Property   Gets or sets the color for all drawn page objects. Use the different color constants for the standard VGA color set. For more information, refer to SigmaPlot Automation from the SigmaPlot Help menu.

Comments Property

Syntax: Notebook/NotebookItems object.Comments

A standard property of notebook files and all NotebookItems objects. Returns or sets the Description field in the Summary Information for all notebook items, or the Comments section under the Summary tab of the Windows 95/98 file Properties dialog box for notebook files.

Count Property

A property available to all collection objects that returns the number of objects within that collection.

CurrentDataItem Property

The CurrentDataItem property returns the worksheet window in focus as an object. You must still use the ActiveDocument property to specify the currently active notebook.

Note that if a worksheet is not in focus an error is returned.

CurrentItem Property

This property returns whatever notebook item currently has focus as an object. You must still use the ActiveDocument property to specify the currently active notebook.

CurrentPageItem Property

Returns the current graph page window as a GraphItem object. You must still use the ActiveDocument property to specify the currently active notebook.

If the current item in focus is not a page, an error is returned.

DataTable Property

Returns the DataTable object for the specified worksheet object.

DefaultPath Property

Sets or returns the default path used by the Application object to save and retrieve files. Files are opened using the Notebooks collection Open method and saved using the Notebook object Save or SaveAs methods.

DropLines Property

Returns the DropLines line collection for a Plot object. Line objects within the DropLines collection have standard line properties.

Use an index to return a specific set of drop lines from the DropLines collection:

1. XYplane (SLA_FLAG_DROPZ , 3D graphs only)

2. Y axis/X direction or YZ plane (SLA_FLAG_DROPX)

3. X axis/Y direction or ZX plane (SLA_FLAG_DROPY)

Some drop line properties are controlled from the Plot object; for example, use the SetAttribute(SLA_PLOTOPTIONS,SLA_FLAG_DROPX Or FLAG_SET_BIT) plot object method to turn on y axis drop lines. Other drop line properties are set using Line object attributes.

Expanded Property

A property of notebook window notebooks and sections, which opens or closes the tree for that notebook section, or returns a true or false value for the current view.

Fill Property   The Fill property is used to return the Solid object for the specified Plot object. Solid objects for plots include bars and boxes.

FullName Property   Returns the filename and path for either the application or the current notebook object. If the notebook object has not yet been saved to a file, an empty string is returned.

Functions Property   The Functions property is used to return the collection of Function objects for the specified Plot object. Plot functions include regression and confidence lines, and all reference (QC) lines. The individual function lines are specified using an index:

Index	Constant	Function
1	SLA_FUNC_REGR	Regression Line
2	SLA_FUNC_CONF1	Upper Confidence Intervals
3	SLA_FUNC_CONF2	Lower Confidence Interval
4	SLA_FUNC_PRED1	Upper Prediction Interval
5	SLA_FUNC_PRED2	Lower Prediction Interval
6	SLA_FUNC_QC1	1st Reference Line (Upper Specification)
7	SLA_FUNC_QC2	2nd Reference Line (Upper Control Line)
8	SLA_FUNC_QC3	3rd Reference Line (Mean)
9	SLA_FUNC_QC4	4th Reference Line (Lower Control Line)
10	SLA_FUNC_QC5	5th Reference Line (Lower Specification)

Note that most regression and reference lines options are controlled with different plot and line attributes. For example, to turn on a regression line, use SetAttribute(SLA_REGROPTIONS,SLA_REGR_FORPLOT Or FLAG_SET_BIT), and to turn on the third reference line, use SetAttribute(SLA_QCOPTIONS,SLA_QCOPTS_SHOWQC3 Or FLAG_SET_BIT)

Graphs Property   Returns the collection of graphs for the specified Page object. Use the index to select a specific Graph object. Graphs are used in turn to return the different graph items: Plots, Axes, the graph title, and the graph legend.

GraphPages Property   Returns the GraphPages collection of Page objects for a GraphItem object. However, since there is currently only one graph page for any given graph item, you can always use GraphPages(0). However, in order to access items within a GraphItem, you must always specify the GraphPage.

Height Property	Sets or returns the height of the application window or specified notebook document window in pixels, or the size of pages and page objects in 1000ths of an inch.
InsertionMode Property	Sets or returns a Boolean indicating whether or not Insert mode is on.
Interactive Property	Sets or returns a Boolean indicating whether or not the user is allowed to interact with the running notebook window or application. Do not set the Application property to False from within SigmaPlot or you will lose access to the application.
IsCurrentBrowser Entry Property	Returns whether or not the specified item is the currently selected item in the notebook tree. This is particularly useful when adding new objects to a notebook in s specific notebook location.
IsCurrentItem Property	Returns whether or not the specified item is the currently selected item. This property is particularly useful when used in conjunction with the CurrentItem property.
IsOpen Property	A property common to all NotebookItems objects. Returns a Boolean indicating whether or not the specified document or section is open. Open and close notebook items using the Open and Close methods.
ItemType Property	A property common to all NotebookItems objects. Returns an integer denoting the item/object type.

1	CT_WORKSHEET	NativeWorksheetItem
2	CT_GRAPHICPAGE	GraphItem
3	CT_FOLDER	SectionItem
4	CT_STATTEST	ReportItem (SigmaStat)
5	CT_REPORT	ReportItem (SigmaPlot)
6	CT_FIT	FitItem
7	CT_NOTEBOOK	NotebookItem
8	CT_EXCELWORKSHEET	ExcelItem
9	CT_TRANSFORM	TransformItem
10		MacroItem

Keywords Property   A standard property of notebook files and all NotebookItems objects. Sets the Keywords field under the Summary tab of the Windows 95/98 file Properties dialog box.

Note that the keywords for notebook items are not currently displayed or used. The default keywords used by SigmaPlot notebooks are "SigmaPlot" and "SigmaStat."

Left Property   Sets or returns the left coordinate of the application window or specified notebook document window in pixels, or the size of pages and page objects in 1000ths of an inch.

Line Property   Returns the Line object for the specified Plot object. Lines are available in both line plots and line and scatter plots.

LineAttributes Property   Returns the collection of axis Line objects for the specified Axis object. Use the collection index to return a specific line object:

Index	Line
1	Axis Lines
2	Major Ticks
3	Minor Ticks
4	Major Grid
5	Minor Grid
6	Axis Break

Note that many axis line attributes are set with the different Axis object attributes, using the Axis object SetAttribute method.

Name Property   A standard property of almost all SigmaPlot objects. Returns or sets the Title name and field in the Summary Information for all notebook items, the filename for a notebook file, and the object name or title for page objects.

To set the title used for a notebook, use the Notebook object Title property, or set the name for NotebookItems(0).

Note: If you attempt to set the name of a document to the existing name, you will receive an error message and the macro will halt.

NamedRanges Property   Returns the collection of NamedDataRanges from a DataTable object. Use the NamedDataRanges collection to return a specific NamedDataRange object.

NameObject Property   Returns the Text object that corresponds to the name of the specified object.

NameOfRange Property	Sets or returns the name for a NamedDataRange object. Useful for returning lists of column and row titles, which are named ranges.

NotebookItems Property	A Notebook object property that returns the collection of notebook items. Use the NotebookItems collection to access individual notebook items. Worksheets, pages, equations, reports, macros, and section and notebook folders are all notebook items and can be returned as objects.

Notebooks Property	An Application object property that returns the Notebooks collection object. Use the Notebooks collection to return individual Notebook objects and create new notebooks.

NumberFormat Property	Sets or returns the format used by the currently selected cells in the DataTable of the NativeWorksheetItem or ExcelItem object. If there is no selection, the format for the entire worksheet is assumed. If there are mixed formats, a NULL value is returned.

Both Number and Date and Time formats are set or returned using the standard number and date and time format designations.

ObjectType Property	Returns the type value for the specified object. The values returned and corresponding object types are:

Value	Constant	Object
1	GPT_PAGE	Page
2	GPT_GRAPH	Graph
3	GPT_PLOT	Plot
4	GPT_AXIS	Axis
5	GPT_TEXT	Text
6	GPT_LINE	Line
7	GPT_SYMBOL	Symbol
8	GPT_SOLID	Solid
9	GPT_TUPLE	Tuple
10	GPT_FUNCTION	Function
11	GPT_EXTERNAL	External
12	GPT_BAG	Group
13	GPT_DATATABLE	DataTable

OwnerGraphObject Property

Returns the object that the current object is contained within. This applies to the different graph page object hierarchies, where the Parent property is not supported.

Parent Property

Returns the object or collection immediately "above" the current object. For graph page items, use the OwnerGraphObject property instead.

Path Property

Returns the default path in which SigmaPlot looks for documents, or the path of the specified notebook file.

For notebooks, you can use the Name property to return the file name without the path, or use the FullName property to return the file name and the path together.

Plots Property

Returns the collection of plots for the specified Graph object. Use an index to return the individual Plot objects for the graph.

Saved Property

Returns a True or False value for whether of not the document has been saved since the last changes. Note that notebook items that are closed from within SigmaPlot are automatically saved to the notebook, but that the notebook file is only saved using a Save or Save As command or method.

SelectedText Property

Returns the text of the current selection from a ReportItem. You can set or return a text selection using the SelectionExtent property.

SelectionExtent Property

Returns the array of current selection extents from a ReportItem or ExcelItem. The start and stop indices for each selection are listed as individual members of the array, e.g., .SelectionExtent(0) is the start of the first selection, and SelectionExtent(1) is the end of the first selection.

ShowStatsWorksheet Property

If this Boolean property is set to "True," SigmaPlot opens up a statistics window that displays statistics about the specified NativeWorksheetItem. Statistics include: mean, standard deviation, standard error, half-widths for 95% and 99% confidence intervals, sample size, total, minimum, maximum, smallest positive value, and number of missing values. If this property is set to "False," the statistics window is closed if open.

This property returns "True" if the statistics worksheet window is open or "False" if the worksheet window is not open or the specified NativeWorksheet is not open.

If the specified NativeWorksheet object is not open, setting this property has no effect.

StatsWorksheetData Table Property

Returns the Column Statistics worksheet as a DataTable object.

Returns an object expression representing the read-only data table belonging to the NativeWorksheetItemís statistics worksheet. If the worksheet has not been opened using the ShowStatsWorksheet property, this property returns nothing.

StatusBar Property

Sets or returns the SigmaPlot application window status bar text. Note that when a macro is running within SigmaPlot, it will also issue status messages that will overwrite messages set with the StatusBar property. A macro running in VB or VBA outside SigmaPlot will not create its own status bar messages other than those set with StatusBar.

StockScheme Property

Returns the property scheme value for a variable, which can then be assigned to a graph object.

STOCKSCHEME_COLOR_BW	&H00010001
STOCKSCHEME_COLOR_GRAYS	&H00020001
STOCKSCHEME_COLOR_EARTH	&H00030001
STOCKSCHEME_COLOR_FOREST	&H00040001
STOCKSCHEME_COLOR_OCEAN	&H00050001
STOCKSCHEME_COLOR_RAINBOW	&H00060001
STOCKSCHEME_COLOR_OLDINCREMENT	&H00070001

STOCKSCHEME_SYMBOL_DOUBLE	&H00010002
STOCKSCHEME_SYMBOL_MONOCHROME	&H00020002
STOCKSCHEME_SYMBOL_DOTTEDDOUBLE	&H00030002
STOCKSCHEME_SYMBOL_OLDINCREMENT	&H00040002

| STOCKSCHEME_LINE_MONOCHROME | &H00010003 |
| STOCKSCHEME_LINE_OLDINCREMENT | &H00020003 |

| STOCKSCHEME_PATTERN_MONOCHROME | &H00010004 |
| STOCKSCHEME_PATTERN_OLDINCREMENT | &H00020004 |

Subject Property

A standard property of notebook files and all NotebookItems objects. Sets the Subject field under the Summary tab of the Windows 95/98 file Properties dialog box.

Note that the Subject for notebook items is not currently displayed or used.

Symbols Property

Returns the Symbol object for the specified Plot object.

Template Property	Returns the Notebook object used as the template source file. The template is used for new page creation. To create a graph page using a template file, use the ApplyPageTemplate method.
Text Property	Specifies the text for the report, transform or macro code. The text is unformatted, plain text.

Σ Use the vbCrLf string data constant to insert a carriage-return and linefeed string.

Transforms: To change the value of a transform variable, use the AddVariableExpression method. Run transforms using the Execute method.

TickLabelAttributes Property	Returns the tick label Text objects for the specified Axis object.
Title Property	A Notebook object property. Sets the Name of the NotebookItem object of the Notebook file, and the Title field under the Summary tab of the Windows 95/98 file Properties dialog box. Does not affect the file name; to change the file name, use either the Name or FullName property.
Top Property	Sets or returns the top coordinate of the application window or specified notebook document window.
Visible Property	A property common to the Application, Notebook, and NotebookItems document objects. Sets or returns a Boolean indicating whether or not the application or specified document window is visible. Do not set the Application property to False from within SigmaPlot or you will lose access to the application.

Note that hidden document windows will still appear in the notebook window tree. Setting Visible=False for a notebook object hides all document windows for the notebook as well.

Width Property	Sets or returns the width of the application window or specified notebook document window.

SigmaPlot Methods

Methods are an action that can be performed on or by an object-think of methods as "verbs." For example, the WorksheetEditItem object has Copy and Clear methods. Methods can have parameters that specify the action ("adverbs").

For more information, refer to SigmaPlot Automation Help from the SigmaPlot Help menu.

Activate Method Makes the specified notebook the object specified by the ActiveDocument property.

Add Method The Add method is used in collections to add a new item to the collection. The parameters depend on the collection type:

Collection	Value	Parameters
Notebooks		None
NotebookItems	1	CT_WORKSHEET
	2	CT_GRAPHICPAGE
	2	CT_FOLDER
	4	CT_STATTEST
	5	CT_REPORT
	6	CT_FIT
	7	CT_NOTEBOOK
	8	CT_EXCELWORKSHEET
	9	CT_TRANSFORMTransformItem
	10	
Graph Objects	2	GPT_GRAPH, more...
	3	GPT_PLOT, more...
	4	GPT_AXIS, more...
	5	GPT_TEXT, more...
	6	GPT_LINE, more...
	7	GPT_SYMBOL, more...
	8	GPT_SOLID, more...
	9	GPT_TUPLE, more...
	10	GPT_FUNCTION, more...
	11	GPT_EXTERNAL, more...
	12	GPT_BAG, more...
NamedRanges		Name string, Left long, Top long, Width long, Height long, NamedRange

The GraphObjects collection uses the CreateGraphFromTemplate and CreateWizardGraph methods to create new GraphObject objects.

AddVariable Expression Method

Allows the substitution of any transform variable with a value.

ApplyPageTemplate Method

Overwrites the current GraphItem using a new page template specified by the template name. Optionally, you can specify the notebook file to use as the source of the template page. If no template file is specified, the default template notebook is used, as returned by the Template property.

AddWizardAxis Method

Adds an additional axis to the current graph and plot on the specified GraphItem object, using the AddWizardAxis options. If there is only one plot for the current graph, SigmaPlot will return an error. Use the following parameters to specify the type of scale, the dimension, and the position for the new axis:

Scale TYPE
SAA_TYPE_LINEAR
SAA_TYPE_COMMON (Base 10)
SAA_TYPE_LOG (Base e)
SAA_TYPE_PROBABILITY
SAA_TYPE_PROBIT
SAA_TYPE_LOGIT

Dimension		
DIM_X	1	The X dimension
DIM_Y	2	The Y dimension
DIM_Z	3	The Z dimension (if applicable)

Position	
AxisPosRightNormal	0
AxisPosRightOffset	1
AxisPosTopNormal	2
AxisPosTopOffset	3

Position	
AxisPosLeftNormal	4
AxisPosLeftOffset	5
AxisPosBottomNormal	6
AxisPosBottomOffset	7

AddWizardPlot Method Adds another plot to the current graph on the specified GraphItem object using the following parameters to define the plot:

Parameter	Values	Optional
graph type	any valid type name	no
graph style	any valid style name	no
data format	any valid data format name	no
column array	any column number/title array	no
columns per plot array	array of columns in each plot	yes
error bar source	any valid source name	error bar plots only
error bar computation	any valid computation name	error bar plots only
angular axis units	any valid angle unit name	polar plots only
lower range bound	any valid degree value	polar plots only
upper range bound	any valid degree value	polar plots only
ternary units	upper range of ternary axis scale	ternary plots only

Clear Method Clears the selection in items that support this.

Close Method The Close method is used to close notebooks and notebook items. The parameters for each object type depend on the object:

Notebook Save before closing Boolean, filename string

NotebookItems Save before closing Boolean

Specifying a Save before closing value of "False" closes the notebook or notebook item without saving changes made to the object.

Note that for NotebookItems and SectionItems, a Close corresponds to an Expanded = False.

| Copy Method | Copies the currently selected item within the specified notebook item. If no item is selected, then an error is returned. |

CreateGraphFrom Template Method Creates a graph for a GraphItem from the Graph Style Gallery.

CreateWizardGraph Method Creates a graph in the specified GraphItem object using the Graph Wizard options. These options are expressed using the following parameters:

Parameter	Values	Optional
graph type	any valid type name	no
graph style	any valid style name	no
data format	any valid data format name	no
columns plotted	any column number/title array	no
columns per plot	array of columns in each plot	yes
error bar source	any valid source name	error bar plots only
error bar computation	any valid computation name	error bar plots only
angular axis units	any valid angle unit name	polar plots only
lower range bound	any valid degree value	polar plots only
upper range bound	any valid degree value	polar plots only
ternary units	upper range of ternary axis scale	ternary plots only

Cut Method Removes the current selection from the specified object, placing the contents on the clipboard. This method is equivalent to using the Copy method, followed by the Clear method. However, whereas Copy places OLE link formats on the clipboard for GraphItem objects, Cut does not.

Delete Method Deletes a notebook item from a NotebookItems collection, as specified using an index number or name. If the item does not exist, an error is returned.

DeleteCells Method Deletes the specified cells from the worksheet. The remaining cells can be moved in two different directions to fill in the deleted region:

1. Shift Cells Up

2. Shift Cells Left

To delete an entire column or row, simply set the column bottom or row right value to the system maximum:

Rows: 32,000,000

Columns: 32,000

Execute Method   Used to execute the specified TransformItem.

Export Method   SigmaPlot Automation supports export of NativeWorksheetItem objects, GraphItem objects, and NotebookItem objects of type CT_NOTEBOOK.

➤ If applied to a NativeWorksheetItem object, this method exports either the data in the worksheet to the specified data format or the entire notebook to a previous SPW file format.

➤ If applied to a GraphItem object, this method exports either the graphic data on the page to the specified graphic format or the entire notebook to a previous SPW file format.

➤ If applied to the first NotebookItem in the NotebookItemList, this method exports the entire notebook to a previous SPW file format.

GetAttribute Method   The GetAttribute method is used by all graph page objects to retrieve current attribute settings. Attributes are numeric values that also have constants assigned to them. For a list of all these attributes and constants, see SigmaPlot Constants.

Message Forwarding: If you use the GetAttribute method to retrieve an attribute that does not exist for the current object, the message is automatically routed to an object that has this attribute using the message forwarding table.

To use the Object Browser to view Constants You can view alternate values for attributes and constants by selecting the current attribute value, then clicking the Object Browser button. All valid alternate values will be listedóto use a different value, select the value and click Paste.

GetData Method   Returns the data within the specified range from a DataTable object as a variant. This variant is always returned as a one dimensional array. If a 2D array is specified, the data is stacked as a linear array. To ensure that GetData retrieves all data in a row or column, specify the worksheet maximum as the right of bottom parameter.

GetMaxLegalSize Method   Initializes the values of the maximum worksheet column and row values, so that they can be returned as a variables.

GetMaxUsedSize Method   Initializes the values of the maximum used worksheet column and row values, so that they can be returned as a variables.

Goto Method Moves worksheet cursor position to the specified cell coordinate for the current NativeWorksheetItem or ExcelItem object.

Help Method Opens an on-line Windows help file to a specific topic context map ID number (as a long) or search index keyword (K-word). You can use either the ID number or an index keyword. If any of the parameters are left empty, the SigmaPlot help file defaults are used.

Import Method Imports a data file with the specified file name into an existing NativeWorksheetItem. You can specify both the import starting location in the SigmaPlot worksheet, as well as the range of data imported.

InsertCells Method Inserts the specified block of cells into the worksheet. The existing cells can be moved in two different directions to accommodate the inserted region:

1. Shift Cells Down

2. Shift Cells Right

To insert an entire column or row, simply set the column bottom or row right value to the system maximum:

Rows: 32,000,000

Columns: 32,000

Mesh Method Converts unsorted xyz triplet data to evenly incremented mesh data, as required by mesh and contour plots. The optional parameters control the results columns, mesh range and increment, and original datapoint weighting. Note that the output columns must be specified if the data is to be returned to the worksheet.

Item Method Returns an object from the collection as specified by the object index number or name. Note that the index begins with 0 by default. The Item method is equivalent to specifying an object from the collection object using an index. If the item does not exist, an error is returned.

ModifyWizardPlot Method Modifies the current plot on the specified GraphItem object using the following parameters:

Parameter	Values	Optional
graph type	any valid type name	no
graph style	any valid style name	no
data format	any valid data format name	no
column array	any column number/title array	no

Parameter	Values	Optional
columns per plot array	array of columns in each plot	yes
error bar source	any valid source name	error bar plots only
error bar computation	any valid computation name	error bar plots only
angular axis units	any valid angle unit name	polar plots only
lower range bound	any valid degree value	polar plots only
upper range bound	any valid degree value	polar plots only
ternary units	upper range of ternary axis scale	ternary plots only

NormalizeTernary Data Method

Normalize three columns of raw data to 100 or 1 for a ternary plot.

Open Method

Opens the notebook specified within the Notebooks collection, or the specified notebook item. The parameter depends upon whether you are opening a notebook or a notebook item.

Note that for NotebookItems and SectionItems, an Open corresponds to an Expanded = True.

Paste Method

Place the contents of the Windows Clipboard into the selected notebook item document, at the current position, if applicable. The format specified is an available clipboard format, as displayed by the Edit menu Paste Special command.

Print Method

Prints the selected item, including any items within specified NotebookItems and SectionItems. Specifying the Notebook prints all items in the notebook.

PrintStatsWorksheet Method

Prints the NativeWorksheetItemís statistics worksheet. If the worksheet has not been opened using the ShowStatsWorksheet property, this method fails.

PutData Method

Places the specified variant into the worksheet starting at the specified location. The data can be a 2D array.

Quit Method

Ends SigmaPlot. If SigmaPlot is in use, then this method is ignored.

Redo Method

Redoes the last undone action for the specified object. If redo has been disabled in SigmaPlot for either the worksheet or page, this method has no effect.

Remove Method

Deletes the specified object. The index can be a number or a name. If the specified index does not exist, an error is returned.

Run Method	Runs a FitItem or Macro without closing the object.
SaveAs Method	Save a notebook file for the first time, or to a new file name and path. Note that you need to provide the file extension. Recognized SigmaPlot notebook file extensions are .JNB, .JNT, and .JFL
Save Method	Saves a Notebook object to disk using the current FullName , or a notebook item to the notebook (without saving the notebook file to disk). If no FullName exists for a notebook, an error occurs. To save a notebook that has not yet been saved, you must use the SaveAs method.
	Note: Transform text can be saved to an .xfm file by naming the transform first with the full file name, extension, and path.
Select Method	Selects all of the items within the specified selection region. In addition, if "Top" equals "Bottom" and "Right" equals "Left," the resulting selection includes the object that the specified point lies within.
	If "AddToSelection" is "False" then the previous selection list is replaced by the new list. If "True," then the newly selected items are added to the existing selection list.
SelectAll Method	Selects the entire contents of the item.
SelectObject Method	Clears the current GraphItem selection list and selects the specified graph object so that it can be altered using the SetSelectedObjectsAttribute method. Line and Solid objects can only be selected if they are top level drawing objects (not child objects of other objects).
SetAttribute Method	The SetAttribute method is used by all graph page objects to change current attribute settings. Attributes are numeric values that also have constants assigned to them. For a list of all these attributes and constants, see SigmaPlot Constants.
	Message Forwarding: If you use the SetAttribute method to change an attribute that does not exist for the current object, the message is automatically routed to an object that has this attribute using the message forwarding table.
	Using the Object Browser to view Constants You can view alternate values for attributes and constants by selecting the current attribute value, then clicking the Object Browser button. All valid alternate values will be listedóto use a different value, select the value and click Paste.
SetCurrentObject Attribute Method	Changes the attribute specified by "Attribute," of the "current object" on the graphics page. Use one of the following three techniques to set the "current object" on the graphics page:
	➤ Click the object using the mouse.

> ➤ Use the SigmaPlot menus (e.g. "Select Graph").
> ➤ Use the SetObjectCurrent method.

If the specified GraphItem is not open or there is no current object of the appropriate type on the page, the method will fail.

SetObjectCurrent Method

Sets the specified object to the "current" object for the purpose of the "SetCurrentObjectAttribute" command.

It the specified GraphItem is not open, the method will fail.

SetSelectedObjects Attribute Method

Changes the attribute specified by "Attribute" for all the selected objects on the graphics page. Select graphics page objects using one of the following two techniques:

> ➤ Click the object with the mouse.
> ➤ Use the SelectObject method.

TransposePaste Method

Pastes the data in the clipboard into the worksheet, transposing the row and column indices of the data such that rows and columns are swapped. If there is nothing in the clipboard or the data is not of the right type, nothing will happen.

Undo Method

Undoes the last performed action for the specified object. If undo has been disabled in SigmaPlot for either the worksheet or page, this method has no effect.

Regression Equation Library

This appendix lists the equations found in the Regression Equation Library.

Polynomial

Linear

$$y = y_0 + ax$$

Quadratic

$$y = y_0 + ax + bx^2$$

Cubic

$$y = y_0 + ax + bx^2 + cx^3$$

Inverse First Order

$$y = y_0 + \frac{a}{x}$$

Inverse Second Order

$$y = y_0 + \frac{a}{x} + \frac{b}{x^2}$$

Inverse Third Order

$$y = y_0 + \frac{a}{x} + \frac{b}{x^2} + \frac{c}{x^3}$$

Peak   Three Parameter Gaussian

$$y = ae^{\left[-0.5\left(\frac{x-x_0}{b}\right)^2\right]}$$

Four Parameter Gaussian

$$y = y_0 + ae^{\left[-0.5\left(\frac{x-x_0}{b}\right)^2\right]}$$

Three Parameter Modified Gaussian

$$y = ae^{\left[-0.5\left(\frac{|x-x_0|}{b}\right)^c\right]}$$

Four Parameter Modified Gaussian

$$y = y_0 + ae^{\left[-0.5\left(\frac{|x-x_0|}{b}\right)^c\right]}$$

Three Parameter Lorentzian

$$y = \frac{a}{1 + \left(\frac{x-x_0}{b}\right)^2}$$

Four Parameter Lorentzian

$$y = y_0 + \frac{a}{1 + \left(\frac{x-x_0}{b}\right)^2}$$

Four Parameter Pseudo-Voigt

$$y = a\left[c\left(\frac{1}{1 + \left(\frac{x - x_0}{b} \right)^2} \right) + (1 - c)e^{-0.5\left(\frac{x - x_0}{b} \right)^2} \right]$$

Five Parameter Pseudo-Voigt

$$y = y_0 + a\left[c\left(\frac{1}{1 + \left(\frac{x - x_0}{b} \right)^2} \right) + (1 - c)e^{-0.5\left(\frac{x - x_0}{b} \right)^2} \right]$$

Three Parameter Log Normal

$$y = ae^{\left[-0.5\left(\frac{\ln\left(\frac{x}{x_0} \right)}{b} \right)^2 \right]}$$

Four Parameter Log Normal

$$y = y_0 + ae^{\left[-0.5\left(\frac{\ln\left(\frac{x}{x_0} \right)}{b} \right)^2 \right]}$$

Four Parameter Weibull

$$y = a\left(\frac{c-1}{c} \right)^{\frac{1-c}{c}} \left[\frac{x - x_0}{b} + \left(\frac{c-1}{c} \right)^{\frac{1}{c}} \right]^{c-1} e^{-\left[\frac{x - x_0}{b} + \left(\frac{c-1}{c} \right)^{\frac{1}{c}} \right]^c} + \frac{c-1}{c}$$

Five Parameter Weibull

$$y = y_0 + a\left(\frac{c-1}{c} \right)^{\frac{1-c}{c}} \left[\frac{x - x_0}{b} + \left(\frac{c-1}{c} \right)^{\frac{1}{c}} \right]^{c-1} e^{-\left[\frac{x - x_0}{b} + \left(\frac{c-1}{c} \right)^{\frac{1}{c}} \right]^c} + \frac{c-1}{c}$$

Sigmoidal   Three Parameter Sigmoid

$$y = \frac{a}{1 + e^{-\left(\frac{x - x_0}{b}\right)}}$$

Four Parameter Sigmoid

$$y = y_0 + \frac{a}{1 + e^{-\left(\frac{x - x_0}{b}\right)}}$$

Five Parameter Sigmoid

$$y = y_0 + \frac{a}{\left[1 + e^{-\left(\frac{x - x_0}{b}\right)}\right]^c}$$

Three Parameter Logistic

$$y = \frac{a}{1 + \left(\frac{x}{x_0}\right)^b}$$

Four Parameter Logistic

$$y = y_0 + \frac{a}{1 + \left(\frac{x}{x_0}\right)^b}$$

Four Parameter Weibull

$$y = a\left[1 - e^{-\left(\frac{x - x_0 + b\ln 2^{\frac{1}{c}}}{b}\right)^c}\right]$$

Five Parameter Weibull

$$y = y_0 + a\left[1 - e^{-\left(\frac{x - x_0 + b\ln 2^{\frac{1}{c}}}{b}\right)^c}\right]$$

Three Parameter Gompertz Growth Model

$$y = ae^{-e^{-\left(\frac{x - x_0}{b}\right)}}$$

Four Parameter Gompertz Growth Model

$$y = y_0 + ae^{-e^{-\left(\frac{x - x_0}{b}\right)}}$$

Three Parameter Hill Function

$$y = \frac{ax^b}{c^b + x^b}$$

Four Parameter Hill Function

$$y = y_0 + \frac{ax^b}{c^b + x^b}$$

Three Parameter Chapman Model

$$y = a(1 - e^{-bx})^c$$

Four Parameter Chapman Model

$$y = y_0 + a(1 - e^{-bx})^c$$

Exponential Decay Two Parameter Single Exponential Decay

$$y = ae^{-bx}$$

Three Parameter Single Exponential Decay

$$y = y_0 + ae^{-bx}$$

Four Parameter Double Exponential Decay

$$y = ae^{-bx} + ce^{-dx}$$

Five Parameter Double Exponential Decay

$$y = y_0 + ae^{-bx} + ce^{-dx}$$

Six Parameter Triple Exponential Decay

$$y = ae^{-bx} + ce^{-dx} + ge^{-hx}$$

Seven Parameter Triple Exponential Decay

$$y = y_0 + ae^{-bx} + ce^{-dx} + ge^{-hx}$$

Modified Three Parameter Single Exponential Decay

$$y = ae^{\left(\frac{b}{x+c}\right)}$$

Exponential Linear Combination

$$y = y_0 + ae^{-bx} + cx$$

Exponential Rise to Maximum

Two Parameter Single Exponential Rise to Maximum

$$y = a(1 - e^{-bx})$$

Three Parameter Single Exponential Rise to Maximum

$$y = y_0 + a(1 - e^{-bx})$$

Four Parameter Double Exponential Rise to Maximum

$$y = a(1 - e^{-bx}) + c(1 - e^{-dx})$$

Five Parameter Double Exponential Rise to Maximum

$$y = y_0 + a(1 - e^{-bx}) + c(1 - e^{-dx})$$

Two Parameter Simple Exponent Rise to Maximum

$$y = a(1 - b^x)$$

Three Parameter Simple Exponent Rise to Maximum

$$y = y_0 + a(1 - b^x)$$

Exponential Growth   One Parameter Single Exponential Growth

$$y = e^{ax}$$

Two Parameter Single Exponential Growth

$$y = ae^{bx}$$

Three Parameter Single Exponential Growth

$$y = y_0 + ae^{bx}$$

Four Parameter Double Exponential Growth

$$y = ae^{bx} + ce^{dx}$$

Five Parameter Double Exponential Growth

$$y = y_0 + ae^{bx} + ce^{dx}$$

Modified One Parameter Single Exponential Growth

$$y = ae^{ax}$$

Modified Two Parameter Single Exponential Growth

$$y = e^{a(x - x_0)}$$

Stirling Model

$$y = y_0 + \frac{a(e^{bx} - 1)}{b}$$

Two Parameter Simple Exponent

$$y = ab^x$$

Three Parameter Simple Exponent

$$y = y_0 + ab^x$$

Modified Two Parameter Simple Exponent

$$y = y_0 + (\log a)a^x$$

Hyperbola   Two Parameter Rectangular Hyperbola

$$y = \frac{ax}{b + x}$$

Three Parameter Rectangular Hyperbola I

$$y = y_0 + \frac{ax}{b + x}$$

Three Parameter Rectangular Hyperbola II

$$y = \frac{ax}{b + x} + cx$$

Four Parameter Double Rectangular Hyperbola

$$y = \frac{ax}{b + x} + \frac{cx}{d + x}$$

Five Parameter Double Rectangular Hyperbola

$$y = \frac{ax}{b + x} + \frac{cx}{d + x} + ex$$

Two Parameter Hyperbolic Decay

$$y = \frac{ab}{b + x}$$

Three Parameter Hyperbolic Decay

$$y = y_0 + \frac{ab}{b + x}$$

Modified Hyperbola I

$$y = \frac{ax}{1 + bx}$$

Modified Hyperbola II

$$y = \frac{x}{a + bx}$$

Modified Hyperbola III

$$y = a - \frac{b}{(1 + cx)^{\frac{1}{d}}}$$

Waveform   Three Parameter Sine

$$y = a\sin\left(\frac{2\pi x}{b} + c\right)$$

Four Parameter Sine

$$y = y_0 + a\sin\left(\frac{2\pi x}{b} + c\right)$$

Three Parameter Sine Squared

$$y = a\left[\sin\left(\frac{2\pi x}{b} + c\right)\right]^2$$

Four Parameter Sine Squared

$$y = y_0 + a\left[\sin\left(\frac{2\pi x}{b} + c\right)\right]^2$$

Four Parameter Damped Sine

$$y = ae^{-\left(\frac{x}{d}\right)}\sin\left(\frac{2\pi x}{b} + c\right)$$

Five Parameter Damped Sine

$$y = y_0 + ae^{-\left(\frac{x}{d}\right)}\sin\left(\frac{2\pi x}{b} + c\right)$$

Modified Sine

$$y = a\sin\left(\frac{\pi(x - x_0)}{b}\right)$$

Modified Sine Squared

$$y = a\left[\sin\left(\frac{\pi(x - x_0)}{b}\right)\right]^2$$

Modified Damped Sine

$$y = ae^{-\left(\frac{x}{c}\right)}\sin\left(\frac{\pi(x - x_0)}{b}\right)$$

Power Two Parameter

$$y = ax^b$$

Three Parameter

$$y = y_0 + ax^b$$

Pareto Function

$$y = 1 - \frac{1}{x^a}$$

Three Parameter Symmetric

$$y = a\left|x - x_0\right|^b$$

Four Parameter Symmetric

$$y = y_0 + a\left|x - x_0\right|^b$$

Modified Two Parameter I

$$y = a(1 - x^{-b})$$

Modified Two Parameter II

$$y = a(1 + x)^b$$

Modified Pareto

$$y = 1 - \frac{1}{(1 + ax)^b}$$

Rational   One Parameter Rational I

$$y = \frac{1}{x + a}$$

One Parameter Rational II

$$y = \frac{1}{1 + ax}$$

Two Parameter Rational I

$$y = \frac{1}{a + bx}$$

Two Parameter Rational II

$$y = \frac{a}{1 + bx}$$

Three Parameter Rational I

$$y = \frac{a + bx}{1 + cx}$$

Three Parameter Rational II

$$y = \frac{1 + ax}{b + cx}$$

Three Parameter Rational III

$$y = \frac{a + bx}{c + x}$$

Three Parameter Rational IV

$$y = \frac{a + x}{b + cx}$$

Four Parameter Rational

$$y = \frac{a + bx}{1 + cx + dx^2}$$

Five Parameter Rational

$$y = \frac{a + bx + cx^2}{1 + dx + ex^2}$$

Six Parameter Rational

$$y = \frac{a + bx + cx^2}{1 + dx + ex^2 + fx^3}$$

Seven Parameter Rational

$$y = \frac{a + bx + cx^2 + dx^3}{1 + ex + fx^2 + gx^3}$$

Eight Parameter Rational

$$y = \frac{a + bx + cx^2 + dx^3}{1 + ex + fx^2 + gx^3 + hx^4}$$

Nine Parameter Rational

$$y = \frac{a + bx + cx^2 + dx^3 + ex^4}{1 + fx + gx^2 + hx^3 + ix^4}$$

Ten Parameter Rational

$$y = \frac{a + bx + cx^2 + dx^3 + ex^4}{1 + fx + gx^2 + hx^3 + ix^4 + jx^5}$$

Eleven Parameter Rational

$$y = \frac{a + bx + cx^2 + dx^3 + ex^4 + fx^5}{1 + gx + hx^2 + ix^3 + jx^4 + kx^5}$$

Logarithm

Two Parameter I

$$y = y_0 + a\ln x$$

Two Parameter II

$$y = a\ln(x - x_0)$$

Two Parameter III

$$y = \ln(a + bx)$$

Second Order

$$y = y_0 + a\ln x + b(\ln x)^2$$

Third Order

$$y = y_0 + a\ln x + b(\ln x)^2 + c(\ln x)^3$$

3 Dimensional   Plane

$$z = z_0 + ax + by$$

Paraboloid

$$z = z_0 + ax + by + cx^2 + dy^2$$

Gaussian

$$z = ae^{-0.5\left[\left(\frac{x-x_0}{b}\right)^2 + \left(\frac{y-y_0}{c}\right)^2\right]}$$

Lorentzian

$$z = \frac{a}{\left[1 + \left(\frac{x-x_0}{b}\right)^2\right]\left[1 + \left(\frac{y-y_0}{c}\right)^2\right]}$$

ndex

Symbols

.FIT files 143
.XFM files 1, 4

A

ABS function 28
Absolute minimum
    sum of squares 221, 222
Accumulation functions 24
Add Procedure Dialog Box 251
Adjusted R^2
    regression results 168
Algorithm
    Marquardt-Levenberg 145, 160, 199
Alpha value
    power 172
ANOVA
    one way ANOVA transform 71—73
ANOVA table
    regression results 169
APE function 28
ARCCOS function 29
ARCSIN function 30
ARCTAN function 30
Area and distance
    functions 25
Area beneath a curve
    transform 73—74
AREA function 31
Arguments, transform 21
    see also function arguments
Arithmetic operators
    transforms 19
Array reference
    example of use 111
Automation 239—252
    introduction 2
    methods 273
    objects 254

properties 264
AVG function 31
Axis scale
    user-defined 131—133
    user-defined transform 131—133

B

Bar chart histogram with Gaussian distribution 111—113
Bivariate statistics transform 74—75
BLOCK function 32
    Fast Fourier transform 94
BLOCKHEIGHT function 32
BLOCKWIDTH function 32

C

Cancelling a regession 162
CELL function 33
CHOOSE function 34
Coefficient of determination
    stepwise regression results 163, 168
Coefficient of determination (R^2) transform 82—83
Coefficient of variation
    parameters 164, 205, 212—215
Coefficients
    regression results 168
COL function 21, 34
Color
    smooth color transition transform 127—129
Comments
    entering regression 186
Completion status messages
    regression results 180—181
COMPLEX function 35
Components
    see transform components 6—9
Computing
    derivatives 87—93
Confidence interval
    linear regressions 113—116
    regression results 174

R